Proceedings in Technology Transfer

The book series "Proceedings in Technology Transfer" provides a valuable platform for researchers, practitioners, and policymakers to share their experiences and insights related to technology transfer. Our aim is to facilitate knowledge exchange between academic researchers and industry professionals, accelerating the introduction of new technologies and driving economic growth.

Our series publishes books and conference proceedings that focus on original research findings, best practice recommendations, emerging trends, and technology evaluations. The series covers all topics which explore the transfer of technology in various industry sectors that require innovations and economic growth.

Join us in advancing the technology transfer and its impact on innovation. Publish your proceedings with us and offer insights and expertise to professionals in your field.

Christina Chin M. M. · Novita Sakundarini ·
Akhil Garg · Andrew Spowage
Editors

Selected Articles from the Smart and Sustainable Industrial Ecosystem Conference 2024

SSIEC 2024, 5 August, Kuala Lumpur, Malaysia

Springer

Editors
Christina Chin M. M.
Department of Mechanical, Materials and Manufacturing Engineering, Faculty of Science and Engineering
University of Nottingham Malaysia
Semenyih, Selangor, Malaysia

Novita Sakundarini
Department of Mechanical, Materials and Manufacturing Engineering, Faculty of Science and Engineering
University of Nottingham Malaysia
Semenyih, Selangor, Malaysia

Akhil Garg
School of Mechanical Science and Engineering
Huazhong University of Science and Technology
Wuhan, Hubei, China

Andrew Spowage
School of Engineering and Materials Science
Queen Mary University of London
London, UK

ISSN 2948-2321 ISSN 2948-233X (electronic)
Proceedings in Technology Transfer
ISBN 978-981-96-2805-6 ISBN 978-981-96-2806-3 (eBook)
https://doi.org/10.1007/978-981-96-2806-3

This Springer imprint is published by the registered company Springer Nature Singapore Pte Ltd.
The registered company address is: 152 Beach Road, #21-01/04 Gateway East, Singapore 189721, Singapore

Preface

The "**Selected Articles from the Smart and Sustainable Industrial Ecosystem Conference**" is a compilation of the insightful and impactful research presented at the SSIEC August 2024 conference in Kuala Lumpur, Malaysia. The event brought together academics to discuss the transformative trends shaping the future of industrial ecosystems. As the global demand for sustainable development grows, industries must adapt to create systems that are both economically viable and environmentally sound. This publication aims to highlight the interdisciplinary approaches, innovative technologies, and strategic policies that contribute to building smart, sustainable industrial ecosystems.

The selected articles in this collection reflect a broad spectrum of research topics, ranging from advancements in green manufacturing, mechanical engineering and circular economy practices and industry applications to digital innovations like artificial intelligence (AI) for optimizing resource use. Additionally, themes of energy efficiency, and sustainable manufacturing technology, blockchain technology are explored, demonstrating the interconnected nature of the challenges and solutions needed for a sustainable industrial future.

The editors would like to extend their gratitude to the contributors and reviewers who played a crucial role in ensuring the quality and relevance of these articles. Their dedication has made this publication a valuable resource for researchers, practitioners, and decision-makers striving to create smarter, more sustainable industrial ecosystems. We hope this compilation not only inspires new ideas and collaborations but also serves as a foundation for future research and innovation in this vital field.

Editors

Acknowledgements

The publication of **"Selected Articles from the Smart and Sustainable Industrial Ecosystem Conference"**would not have been possible without the support, dedication, and collaboration of numerous individuals. First and foremost, we would like to express our sincere appreciation to all the authors who contributed their research and shared their expertise, making this compilation a valuable contribution to the field of sustainable industrial ecosystems.

We are deeply grateful to the conference organizing committee for their meticulous planning and coordination, which ensured a platform for fruitful discussions and knowledge exchange. Special thanks are due to the reviewers, whose time, effort, and insightful feedback helped maintain the high standard of the selected articles. Their contributions were instrumental in enhancing the quality and rigor of this publication.

We would also like to acknowledge the support of our partner institutions, whose commitment to advancing sustainability and smart industrial practices made this conference and subsequent publication a reality. Their encouragement and resources were pivotal in bringing together a diverse community of researchers and practitioners dedicated to building a more sustainable future.

Finally, we extend our gratitude to the editorial team for their dedication, professionalism, and unwavering support throughout the publication process. Their collaborative efforts ensured that each article was presented in its best possible form.

To all who contributed to this endeavour, we offer our heartfelt thanks. We hope this collection serves as a meaningful resource and inspires continued innovation and progress toward sustainable industrial ecosystems.

Editors

About This Book

The book titled **"Selected Articles from the Smart & Sustainable Industrial Ecosystem Conference 2024"** is a comprehensive collection of scholarly articles presented at the 2024 edition of the conference. This conference brought together experts, researchers, and practitioners in the fields of industrial sustainability and smart ecosystems to share insights, innovations, and breakthroughs in these areas. The book is a vital resource for those interested in the intersection of technology, sustainability, and industrial development. The selected articles delve into key topics such as:

Sustainable Industry Practices: Articles focus on innovative methods for reducing the environmental impact of industrial processes, from waste reduction and recycling to energy efficiency and resource optimization.

Smart Manufacturing: Several papers discuss the integration of advanced technologies, such as the biometric system, artificial intelligence (AI), autonomous mobile and into manufacturing systems to enhance productivity while maintaining sustainability goals.

Circular Economy Models: This section highlights how industries are transitioning from linear production models to circular ones, emphasizing reuse, recycling, and the reduction of waste. Articles here explore case studies, and agricultural waste application to promote a circular economy.

The book serves as a valuable reference for academics, industry professionals, and students who are keen on understanding the latest trends in creating sustainable industrial ecosystems. Each article offers a blend of theoretical insights and practical applications, making it relevant for those working to implement smart and sustainable strategies in their organizations.

Keywords: Sustainability · Industrial · Ecosystem · Smart · Additive manufacturing · Autonomous · Innovation

Contents

About the Editors

Ts. Dr. Christina Chin M. M. is an Associate Professor with the University of Nottingham Malaysia in the Department of Mechanical, Materials & Manufacturing Engineering. She is also the departmental Undergraduate Course Director overseeing, managing and providing strategic lead all the undergraduates programmes. She has a wide range multi and interdisciplinary knowledge in project engineering management for sustainability & environmental impact industries, energy management related aspect including green technology, energy trading, security and storage capacity and efficiency. She has earned her Professional Technologist (Ts.) and Graduate Technologist from Malaysia Board of Technology accrediting her skills and experiences. With many publications to date in internationally refereed journals, technical papers, congress/conference proceedings, and book chapters, she is also an active reviewer for international journals, conferences, books and invited guest speakers for lectures and webinars based on well-established academic-industrial collaboration partnership in various countries globally.

Dr. Novita Sakundarini has her core research competencies in eco-design, sustainable manufacturing, material selection, recycling oriented design, design and development of environmental impact assessment tool in product design, and development of expert systems for green design. She established many research collaboration and university linkage in the area of manufacturing and sustainable design, specifically Asia-Pacific universities. She published research articles in many high impacts referred journals, proceeding of international conference and serves as an active invited reviewer of several international journals. She is a Member of the Institution Engineering Designers (IED) UK and Fellow of the Higher Education Academy UK. Apart of her teaching and research activities, she currently serves as Mechatronic Liaison and Benchmarking officer at M3 Department.

Dr. Akhil Garg obtained Bachelor Degree in Mechanical Engineering from National Institute of Technology, Rourkela in the May 2010. In August 2014, he received a doctorate degree in mechanical and aerospace engineering from Nanyang Technological University, Singapore. From September 2014 to March 2016, he was engaged in postdoctoral research in the School of Engineering, Nanyang Technological University, Singapore. From March 2016 to August 2016, he worked as a lecturer in Shantou University in School of Mechatronics Engineering. From September 2016 to July 2019, he worked as an Associate Professor at Shantou University. In August 2019, he joined School of Mechanical Science and Engineering at Huazhong University of Science and Technology. The main research directions include the application of evolutionary algorithm and artificial intelligence in engineering design or design optimization, such as energy storage system design. In the past 5 years, SCI papers has been published in International Journals such as Structural and Multidisciplinary Optimization, Swarm and Evolutionary

Computation, Applied Soft Computing, Expert Systems with Applications, Engineering Applications of Artificial Intelligence, Journal of Intelligent Manufacturing, Journal of Power Sources, Energy and other journals. 96 papers, including 2 ESI hot papers and 10 ESI highly cited papers, H index 21. He presided over the project of "sailing plan" in Guangdong Province to introduce the top talents in short supply. He is currently guest editor of SCI Journal "Journal of Intelligent Manufacturing" and served as a reviewer of more than 30 SCI journals.

Dr. Andrew Spowage is an academic leader with over 25 years of experience in global higher education and senior positions in industry. Offering expertise in senior management, teaching, applied research, student-centred pedagogies and internationalising higher education. He has had a diverse career spanning both leading academic institutions as well as senior roles in a wide range of industrial sectors. He currently holds the position of Reader within the School of Engineering and Materials Science at Queen Mary University of London and Director (Industry) Centre for Research in Engineering and Materials Education. His main teaching and administrative responsibilities focus on the schools' programmes in Xi'an, China. Before joining QMUL, he led the Materials, Metallurgy, Corrosion, Integrity and Failure Investigation teams for Wood PLC across Asia. Prior to joining Wood, he was the Head of the Department of Mechanical, Materials and Engineering, Director of Studies and Associate Professor for the University of Nottingham. During his time, he helped build the university's Malaysia campus and develop its overseas teaching operations.

Hydrothermal Ageing of Copper Oxide Doped Alumina Toughened Zirconia

Mohamed Abbas[1(✉)], Ramesh Singh[2], and Alaa Elsafi Ahmed[3]

[1] Center for Advanced Materials, Qatar University, P.O. Box 2713, Doha, Qatar
m.kamal@qu.edu.qa

[2] Center of Advanced Manufacturing and Material Processing, Department of Mechanical Engineering, Faculty of Engineering, University of Malaya, 50603 Kuala Lumpur, Malaysia

[3] Materials Science and Technology Program, Department of Mathematics, Statistics and Physics, College of Arts and Sciences, Qatar University, P.O. Box 2713, Doha, Qatar

Abstract. This study investigates the hydrothermal aging behavior of undoped and copper oxide-doped alumina-toughened zirconia (ATZ). The ATZ ceramic composites underwent conventional sintering at temperatures ranging from 1250 to 1500 °C with a holding time of 12 min. XRD analysis revealed a stable 100% tetragonal phase for sintered ATZ samples up to 1450 °C, even after 100 h of exposure. At 1500°C, XRD patterns of both undoped and doped ATZ samples showed no phase transformation after up to 3 h of exposure to superheated steam. Extended exposure, however, resulted in phase transformation beyond 10 h. CuO-doped ATZ samples initially exhibited lower monoclinic content, gradually increasing with aging. Undoped ATZ demonstrated better aging resistance, maintaining ~ 40% monoclinic content after 100 h. FESEM images post aging revealed surface roughness changes due to the tetragonal-to-monoclinic phase transformation, with limited nucleation in the largest tetragonal grains. Fracture analysis exhibited macrocracks and microcracks on the transformed surface layer after aging. This study found that 0.2wt% CuO doping did not prevent the low-temperature degradation (LTD) phenomenon at elevated temperature.

Keywords: Alumina Toughened Zirconia · Conventional Sintering · Copper Oxide · Hydrothermal Ageing

1 Introduction

In recent decades, ceramic materials have become integral in both biomedical and structural applications. Among these, zirconia (ZrO_2) has emerged as a highly promising ceramic, leading the field due to its exceptional properties. These include bioinert characteristics, high hardness, impressive thermal shock resistance, corrosion, and fracture resistance, along with superior toughness [1–3]. The success of yttria-stabilized zirconia (Y_2O_3-ZrO_2) in producing dense bodies with remarkable mechanical properties has paved the way for its widespread applications. Known as Yttria-Stabilized Tetragonal Zirconia Polycrystal (Y-TZP), this monolithic ceramic is composed of tetragonal equiaxed grains partially stabilized with 3 mol% yttria [4]. Y-TZP exhibits excellent mechanical properties, biocompatibility, and wear resistance, making it suitable

Chin, C. M. M. et al. (Eds.): SSIEC 2024, PTT, pp. 1–7, 2025.
https://doi.org/10.1007/978-981-96-2806-3_1

for diverse applications. The mechanical improvement in Y-TZP is attributed to stress-induced tetragonal-to-monoclinic phase transformation, known as transformation toughening [5, 6]. This process results in a volume increase of 3–5% that inhibits crack propagation and increases material toughness. However, this same phase transformation is also responsible for the spontaneous deterioration of mechanical properties, known as hydrothermal aging or low-temperature degradation (LTD), especially when exposed to humid environments at temperatures between 100 and 300 °C [7]. Several studies have demonstrated that incorporating small amounts of sintering additives can enhance the densification of Y-TZP at low temperatures. This, in turn, allows for the control of microstructure, limiting the progression of LTD and contributing to the overall improvement of the mechanical properties of the sintered material [8, 9]. The combination of 20 wt% Al_2O_3 and 80 wt% 3Y-TZP, resulting in a ceramic composite referred to as alumina-toughened zirconia (ATZ), represents one of the most widely studied compositions [10]. The inclusion of 20 wt% alumina (Al_2O_3) in Y-TZP was initially observed to be advantageous in reducing the formation of the monoclinic phase during the aging process. However, a recent study by [11] found that the addition of Al_2O_3 to Y-TZP was not beneficial in suppressing LTD. Their research revealed that ATZ sintered at low temperatures was resistant to hydrothermal aging, while ATZ sintered at high temperatures remained prone to degradation. This discrepancy arises because larger tetragonal grains in high-temperature sintered ATZ are more susceptible to the destabilizing effects of hydroxyl ions during aging. These ions can create residual stresses in the zirconia lattice, potentially triggering a phase transformation that leads to decreased strength and toughness. To address these challenges, researchers have suggested that the mechanical properties of Y-TZP can be preserved and enhanced by regulating hydrothermal aging through the incorporation of specific dopants or precise amounts of sintering additives [9, 12]. Our previous research [11] has focused on optimizing the composition and processing parameters of ATZ to enhance its resistance to hydrothermal aging. We investigated the effects of various sintering temperatures and alumina content on the microstructure and phase stability of ATZ. These studies revealed the complex interplay between processing conditions, microstructure, and hydrothermal aging behavior, highlighting the need for further exploration of novel approaches to mitigate LTD in zirconia-based ceramics. Building upon these findings, the current study extends our line of inquiry by investigating the potential of copper oxide (CuO) as a dopant in ATZ systems. CuO has shown promise in other ceramic systems for improving densification and mechanical properties, but its effects on the hydrothermal aging of ATZ remain largely unexplored [2]. Therefore, this study aims to investigate the impact of CuO doping on the hydrothermal aging behavior of Alumina Toughened Zirconia.

2 Methodology

This study utilized a commercial composite powder (Kyoritsu, Japan) consisting of 3 mol% yttria-stabilized tetragonal zirconia polycrystal (3YTZP) and alumina (Al_2O_3) as the raw materials. The ATZ composite powder was produced by mixing 80 wt% 3YTZP and 20 wt% Al_2O_3 through attrition milling. To prevent the high-surface-energy alumina molecules from agglomerating, zirconia balls with a diameter of 3 mm were employed as milling media, with ethanol as the mixing medium. Subsequently, the slurries were dried in an oven at 60 °C for 24 h. The resulting dried powders were crushed and sieved through a 212 μm sieve to obtain ATZ powders. The ATZ powder was further mixed with 0.2 wt% copper oxide (99.99%, Sigma-Aldrich, USA) using the aforementioned method to form doped samples. Based on the outcome of an earlier study, it was revealed that the 0.2 wt% CuO doped ATZ demonstrated superior densification and outstanding mechanical properties compared to other doped samples [2]. 2.5 g of the blended powders were compacted into a cylindrical steel die cavity with a 20 mm diameter through uniaxial pressing at a pressure of 4 MPa to produce undoped and doped ATZ pellets. Following this, the pellets underwent cold isostatic pressing at 200 MPa for 1 min to achieve a consistent density in the compacted powders. The green specimens underwent conventional sintering in a box furnace at varying temperatures, specifically 1250 °C, 1350 °C, 1450 °C, and 1500 °C, for a duration of 12 min with a consistent heating and cooling rate of 10 °C/min. Considering the thermally activated tetragonal to monoclinic transformation, influenced by the presence of water or water vapor, an ageing experiment was conducted under autoclave conditions with distilled water heated to 180 °C at 10 bar pressure, for durations extending up to 100 h. The impact of hydrothermal ageing on phase stability and microstructural characteristics was assessed using XRD and FESEM techniques, respectively.

3 Results and Discussion

It is well documented that the LTD of tetragonal zirconia is a thermal-activated process and accelerated by the presence of water or water vapour (Ramesh et al., 2018). According to the ISO 13356 (2008) that an accelerated test to simulate the actual LTD process can be conducted under autoclave condition containing steam at elevated temperatures. It has also been suggested by several researchers [13, 14] that 1 h of exposure in such autoclave condition represents about 4 years of ageing in a human under body temperature.

Prior to the hydrothermal aging experiment, both undoped and doped ATZ samples underwent XRD investigation, revealing that all samples exhibited a 100% tetragonal phase, regardless of the sintering temperatures. Thus, no monoclinic phase was present even after sintering at 1500 °C. Moreover, the XRD analysis conducted on sintered ATZ samples indicated that the stability of the tetragonal phase was not disrupted for samples sintered up to 1450 °C, even after 100 h of exposure. On the other hand, the XRD patterns of the aged undoped and doped ATZ samples show no phase transformation for all samples after exposure to superheated steam for periods up to 3 h. However, with a more extended exposure to superheated steam, the aging-induced phase transformation becomes evident when the aging period reaches 10 h and beyond, as shown in Fig. 1.

The monoclinic content of the Cu-ATZ was initially lower than that of ATZ. XRD analyses of the undoped and doped ATZ sintered samples revealed an initial monoclinic phase content of approximately 10%. With further aging, the doped ATZ experienced a sharp rise in monoclinic content, reaching about 41%, compared to undoped ATZ with a monoclinic content of ~27% after 24 h of aging. Consequently, the undoped ATZ sample demonstrated better overall aging resistance, experiencing a monoclinic content

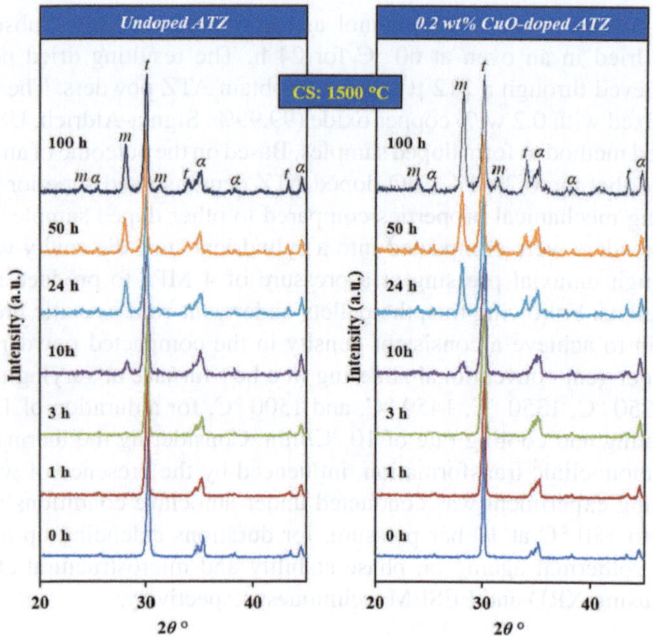

Fig. 1. XRD analysis of aged undoped and doped ATZs sintered at 1500 °C.

of ~40% after 100 h of aging, compared to ~56% for Cu-ATZ. While the addition of 0.2 wt% CuO to ATZ resulted in a lower initial monoclinic content, it did not suppress the aging-induced phase transformation over extended periods. In fact, Cu-ATZ showed a more rapid increase in monoclinic content during aging. These findings indicate that adding CuO to ATZ did not improve long-term resistance to hydrothermal aging under the conditions tested.

Figure 2 presents FESEM images of samples aged for 100 h under superheated steam conditions (180 °C/10 bar). While grain boundaries are distinctly visible, identifying specific aging effects on the sintered sample surfaces proved challenging. The FESEM analysis did not reveal significant microstructural damage following accelerated aging. Nevertheless, the tetragonal to monoclinic phase transformation did alter the surface roughness of the grains [11]. Interestingly, despite XRD analysis confirming the presence of the monoclinic phase, FESEM images showed that monoclinic nuclei formation during aging was confined to the grain boundary regions of the largest tetragonal grains in the Y-TZP structure. This localized transformation is likely due to the addition of alumina and copper oxide, which influence the tetragonal to monoclinic transformation under hydrothermal conditions, particularly in larger tetragonal grains [11]. Notable in these images is the presence of an exaggerated coarse texture, with alumina grains occasionally exhibiting elongated morphologies.

FESEM analysis of the fractured surfaces of ATZ samples after 100 h of aging exposure revealed significant microstructural changes, as shown in Fig. 3. The tetragonal-to-monoclinic phase transformation resulted in the formation of macro-and microcracks on the transformed surface layer [11]. It is evident that the doped ATZ sample, aged and

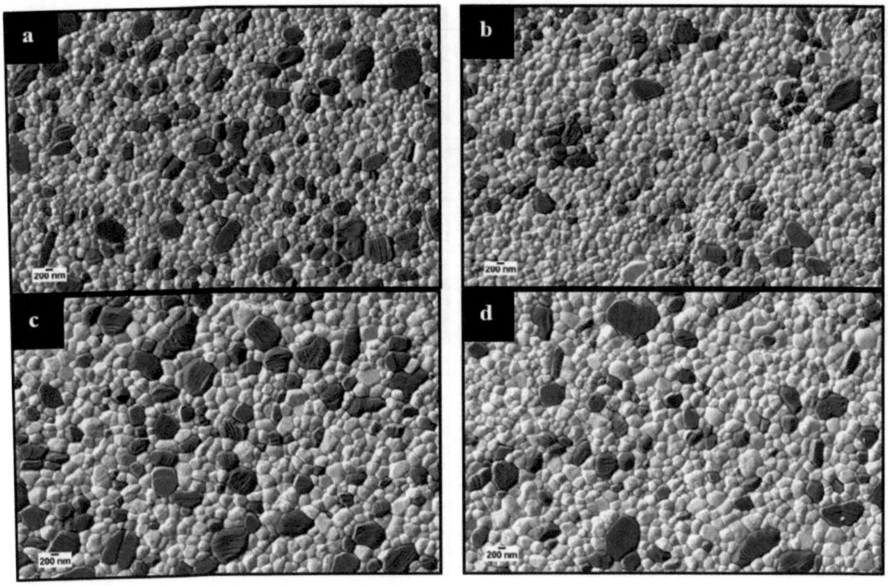

Fig. 2. SEM images of sintered samples after 100 h of ageing (a) Cu-ATZ sintered at 1450 °C/12 min. (b) undoped ATZ sintered at 1450 °C/12 min. (c) Cu-ATZ sintered at 1500 °C/12 min. (d) undoped ATZ sintered at 1500 °C/12 min.

sintered at 1450 °C (Fig. 3a), exhibited a combination of intergranular and transgranular fracture modes. In contrast, the fracture mode of the undoped ATZ aged sample (Fig. 3b) was predominantly intergranular, characterized by the presence of holes on the fracture surface, indicative of intergranular fracture morphology. This intergranular mode allows for easier identification of tetragonal grains or newly transformed monoclinic grains and their morphology. Pits were observed on the fractured surfaces of aged ATZ samples, particularly in the undoped ATZ. This suggests that the undoped ATZ may be in the early stages of monoclinic phase transformation. It is anticipated that extended aging would lead to increased monoclinic content [11].

For samples sintered at 1500 °C, the doped ATZ (Fig. 3c) showed a higher proportion of transgranular fracture. The undoped ATZ (Fig. 3d) exhibited a mixed fracture mode with numerous pits on the fractured surface, though the presence of grains with sharp edges indicated a predominance of intergranular fracture. The prevalence of intergranular fracture is likely due to a combination of internal pressures from the tetragonal-to-monoclinic phase transformation and weakened intergranular bonds resulting from water diffusion [11, 15]. These observations provide insight into the aging mechanisms and their effects on the microstructural integrity of ATZ ceramics.

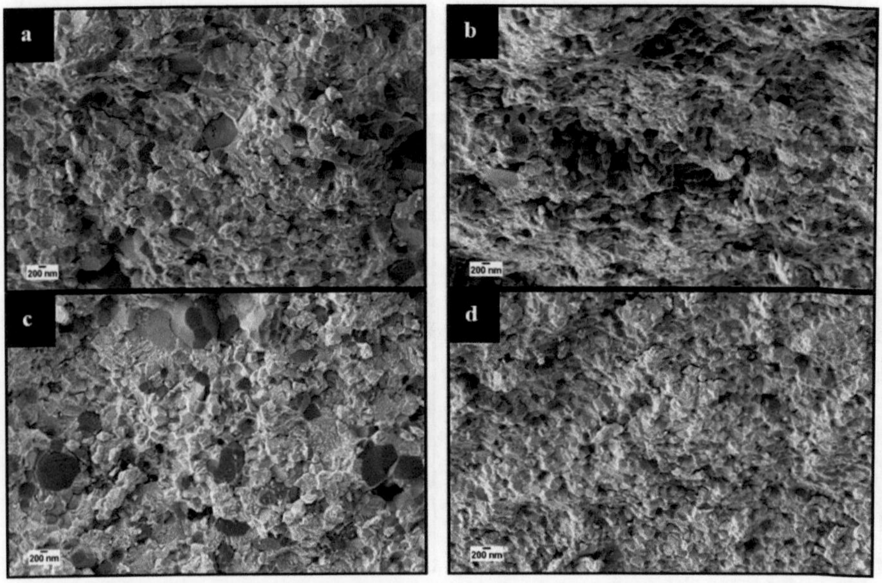

Fig. 3. SEM micrographs showing the fracture surfaces of sintered samples after 100 h of ageing (a) Cu-ATZ sintered at 1450 °C/12 min. (b) undoped ATZ sintered at 1450 °C/12 min. (c) Cu-ATZ sintered at 1500 °C/12 min. (d) undoped ATZ sintered at 1500 °C/12 min.

4 Conclusion

This study investigated the hydrothermal aging behavior of undoped and copper oxide-doped alumina-toughened zirconia (ATZ) under various sintering conditions and aging periods. The research provides valuable insights into the effects of CuO doping on the low-temperature degradation (LTD) phenomenon in ATZ ceramics. Key findings indicate that ATZ samples sintered up to 1450 °C retained a stable 100% tetragonal phase, even after prolonged hydrothermal exposure, highlighting the robustness of the ATZ system under moderate sintering conditions. However, samples sintered at 1500 °C exhibited a time-dependent onset of phase transformation after extended exposure to superheated steam. The addition of 0.2 wt% CuO to ATZ initially reduced monoclinic content but did not prevent aging-induced phase transformation over time. In fact, Cu-ATZ samples showed a more rapid increase in monoclinic content during aging compared to undoped ATZ, with the latter demonstrating superior long-term aging resistance. Microstructural analysis revealed phase transformation primarily at grain boundaries, contributing to surface roughness changes, and a mix of fracture modes, suggesting grain boundary weakening due to water diffusion and internal pressures. Ultimately, while CuO doping initially reduced monoclinic content, it did not prevent the LTD phenomenon at elevated temperatures over extended periods. This study emphasizes the complex interplay between composition, processing conditions, and aging behavior in ATZ ceramics, and suggests the need for further research into alternative dopants or concentrations to enhance long-term hydrothermal aging resistance, particularly for applications involving extended exposure to humid environments at elevated temperatures.

Acknowledgement. This research was supported by Qatar University Student Grant **'QUST-1-CAM-2024-1694'**. The findings achieved herein are solely the responsibility of the authors.

References

1. Maziero Volpato, C.A., Altoe Garbelotto, L.G.D., Celso, M., Bondioli, F.: Application of zirconia in dentistry: biological, mechanical and optical considerations. In: Advances in Ceramics - Electric and Magnetic Ceramics, Bioceramics, Ceramics and Environment, pp. 397–420 (2011). https://doi.org/10.5772/21630
2. Abbas, M.K.G., et al.: Densification of copper oxide doped alumina toughened zirconia by conventional sintering. Ceram. Int. **48** (2022). https://doi.org/10.1016/j.ceramint.2021.11.171
3. Kelly, J.R., Denry, I.: Stabilized zirconia as a structural ceramic: an overview. Dent. Mater. **24**, 289–298 (2008). https://doi.org/10.1016/j.dental.2007.05.005
4. Abbas, M.K.G., Ramesh, S., Tasfy, S.F.H., Lee, K.Y.S., Gul, M., Aljaoni, B.: Effect of sintering additives on the properties of alumina toughened zirconia (ATZ). MRS Commun. 1–9 (2023). https://doi.org/10.1557/S43579-023-00400-Y/FIG.S/4
5. Ramesh, S., et al.: Effect of microwave sintering on the properties of copper oxide doped Y-TZP ceramics. Ceram. Int. **44**, 19639–19645 (2018). https://doi.org/10.1016/j.ceramint.2018.07.215
6. Ramesh, S., Sara Lee, K.Y., Tan, C.Y.: A review on the hydrothermal ageing behaviour of Y-TZP ceramics. Ceram. Int. **44**, 20620–20634 (2018). https://doi.org/10.1016/j.ceramint.2018.08.216
7. Abbas, M.K.G., et al.: Effects of sintering additives on the densification and properties of alumina-toughened zirconia ceramic composites. Ceram. Int. **46**, 27539–27549 (2020). https://doi.org/10.1016/j.ceramint.2020.07.246
8. Abbas, M.K.G., Ramesh, S., Tasfy, S.F.H., Lee, K.Y.S.: Effect of microwave sintering on the properties of copper oxide doped alumina toughened zirconia (ATZ). J. Market. Res. **25**, 1041–1054 (2023). https://doi.org/10.1016/J.JMRT.2023.06.014
9. Presenda, Á., Salvador, M.D., Moreno, R., Borrell, A.: Hydrothermal degradation behavior of Y-TZP ceramics sintered by nonconventional microwave technology. J. Am. Ceram. Soc. **98**, 3680–3689 (2015). https://doi.org/10.1111/jace.13810
10. Abbas, M.K.G., Ramesh, S., Tasfy, S.F.H., Lee, K.Y.S.: A state-of-the-art review on alumina toughened zirconia ceramic composites. Mater. Today Commun. **37**, 106964 (2023). https://doi.org/10.1016/J.MTCOMM.2023.106964
11. Abbas, M.K.G., Ramesh, S., Wu, T., Sutharsini, U., Bang, L.T.: Hydrothermal ageing of alumina-toughened zirconia. AIP Conf. Proc. **2643**, 050018 (2023). https://doi.org/10.1063/5.0110660
12. Camposilvan, E., et al.: Aging resistance, mechanical properties and translucency of different yttria-stabilized zirconia ceramics for monolithic dental crown applications. Dent. Mater. **34**, 879–890 (2018). https://doi.org/10.1016/j.dental.2018.03.006
13. Chevalier, J., Cales, B., Drouin, J.M.: Low-temperature aging of Y-TZP ceramics. J. Am. Ceram. Soc. **82**, 2150–2154 (1999). https://doi.org/10.1111/j.1151-2916.1999.tb02055.x
14. Yang, H., Xu, Y.L., Hong, G., Yu, H.: Effects of low-temperature degradation on the surface roughness of yttria-stabilized tetragonal zirconia polycrystal ceramics: a systematic review and meta-analysis. J. Prosthet. Dent. **125**, 222–230 (2021). https://doi.org/10.1016/J.PROSDENT.2020.01.005
15. Kohorst, P., et al.: Low-temperature degradation of different zirconia ceramics for dental applications. Acta Biomater. **8**, 1213–1220 (2012). https://doi.org/10.1016/j.actbio.2011.11.016

Analysis of Product Genes and Symbols of Wooden Furniture—A Case Study of a Brand Design

Chun-Kuan Wu(✉), Ming-Yu Hsiao, Hsuan Lin, and Yann-Long Lee

Department of Industrial Design, Chaoyang University of Technology, Taichung City, Taiwan
jkwu3612@gmail.com

Abstract. Furniture products are essential to human life, carrying cultural heritage and human history. Wooden furniture can showcase the characteristics and evolutionary processes of different eras. Although furniture continuously updates and changes, it contains genetic features similar to biological inheritance, with relationships influencing each other across different periods, making it a valuable subject for exploration. In this study, these furniture characteristics are transformed into product genes and product symbols for analysis. Using the product symbol triangle method, the triadic relationship of symbol representation, object, and interpretation is analyzed. Additionally, the genealogical and adjacent axes methods are combined to analyze design cases of well-known Taiwanese wooden furniture brands. Through these methods, the sources of product features and product genes are analyzed, explaining how these features transform into product representations and establish the sequence relationship of product forms and functions. By summarizing the design styles of products, the relationship between product symbols and product genes can be identified, providing a basis for considering product design direction for the industry when building brands.

Keywords: Product Semiotic Triangle · Paradigmatic Axis · Syntagmatic Axis · Product Styling

1 Introduction

1.1 Background

Furniture products are essential commodities in the development of human civilization, carrying the inheritance of culture and recording the history of humanity. Wooden furniture can better demonstrate the characteristics and evolution of different eras from Jian, R. (2000). Despite continuous updates and changes in furniture design, genetic features exist akin to those found in biological inheritance, establishing relationships of mutual influence across different periods.

In this study, the author (2023) used the product symbol triangle to analyze the symbolic representations, objects, and interpretations of wooden furniture products. These symbolic representations form the product representations, and by examining the

Chin, C. M. M. et al. (Eds.): SSIEC 2024, PTT, pp. 8–14, 2025.
https://doi.org/10.1007/978-981-96-2806-3_2

form in which these symbols are presented, the relationship between the representations, objects, and interpretations within the product symbol triangle is organized. This analysis utilizes the processes of encoding and decoding in semiotics.

For example, as shown in the Fig. 1, what characteristics of a chair do designers and consumers or users think of in the context of the object? What shapes or functions does the chair have in terms of object characteristics? What elements does the designer choose in the encoding process? How does the designer encode these elements? What characteristics (genes) do consumers or users have when decoding these elements (carriers)? How do they decode the symbolic representations of furniture products through interpretation? Do these product representations or product genes have a brand image relationship with the brand? These questions will be analyzed through the research methods.

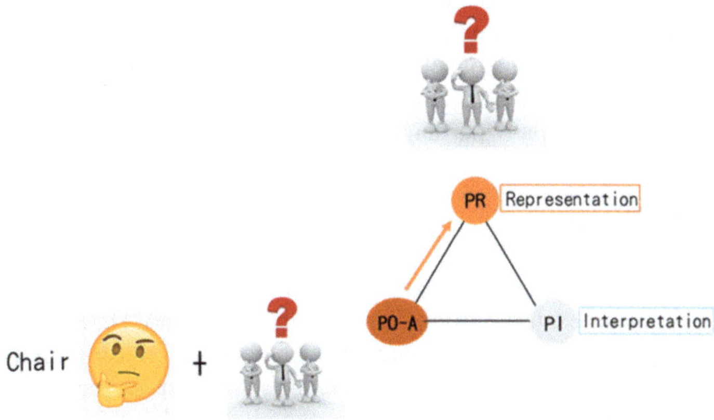

Fig. 1. Diagram of the Product Symbol Triangle Applied to Chair Design

This study applies the product semiotic triangle from Hsiao, M.-Y's (2014) and combines the paradigmatic axis and syntagmatic axis research methods. Ultimately, it summarizes the design style of products and identifies the relationship between product symbols and product genes. The aim is to provide a basis for thinking about product design directions for the industry when building a brand.

2 Methods

This study applies Hsiao, M.-Y's (2014) concept of the product symbol triangle, which encompasses the three realms of Pierce's (1998) semiotics: semiotics, semantics, and pragmatics, to discuss the themes of symbol composition, image representation, and meaning interpretation, respectively.

Regarding product semiotics, the study focuses on architectural, furniture, and home decor imagery symbols as the primary research subjects. Regarding product semantics, it examines the relationship between formal elements and the object of reference through the example of Research Case furniture. Product pragmatics, on the other hand, revolves around design interpretation, including the contextual aspects of symbols and the interpretations of these symbol representations by designers and users.

The research converts these furniture features into product genes and product symbols for analysis. Utilizing the method of the product symbol triangle, the study analyzes the triadic relationship of representation-object-interpretation. Saussure proposed that there are two ways in which signs are composed: the paradigmatic axis and the syntagmatic axis (Terence Hawkes, 1977; T. Hawkes, 1989). The paradigmatic axis involves the intention of selection, while the syntagmatic axis is a concept of organization (Fiske, 1990). After analyzing the product symbols using the triangular model, methods involving both the paradigmatic and syntagmatic axes were combined to analyze design cases of a well-known Taiwanese wooden furniture brand.

The study analyzes the origin of product characteristics and product genes through the aforementioned methods, explains how these features are transformed into product representations and establishes the sequence relationships of product form and function. Research was started with listing of selection criteria of exoskeleton. After the 12 selection criteria has been list out, the criteria were inserted to AHP method with expert from automotive production company participated in the research.

3 Result and Discussion

The Using Danish designer Hans J. Wegner's classic CH24, also known as the Wishbone Chair or Y-Chair, as an example from Chai, C., Shen, D., Bao, D., & Sun, L. (2018). The designer simplifies the observed object through observation, forming the representation of the chair's shape, making the chair resemble the letter Y. According to product semiotics, through the concrete shaping rules, the chair is the object in the semiotic triangle, and the letter Y is the carrier used for interpretation (PI). After interpretation, the designer uses the Y shape on the backrest to form the chair's representation (PR). Therefore, the design element of the chair is encoded in a form similar to the uppercase letter Y, and users, based on their life experiences, recognize the clear identifying symbol of the letter Y in the shape. Hence, they decode the product using the lines of the Y shape, and this chair is commonly referred to as the Y-Chair in the market.

In the research case, the interpretation of the product symbol triangle using the example of a minimalist series round-back chair (Fig. 2) from Zhuang, M., & Xiao, K. (2009) mentions that in Ming-style chairs, the round-back chair is a uniquely styled seating option. It is named for its circular backrest, which extends smoothly from the back to the front, forming the armrests on both sides, giving it a modern feel. In this case, the object refers to the theme of describing Chinese furniture, specifically the "round-back chair." The designer reinterprets the traditional Chinese round-back chair as modern Oriental furniture, and the case brand aims to become synonymous with new Oriental furniture in Taiwan. The term "new Oriental furniture" is used to connect traditional Chinese furniture with modern Oriental furniture. It can be deduced that the brand uses rhetoric to reinterpret the traditional Chinese round-back chair's symbolic representation in modern Oriental furniture. In the product symbol triangle, the Chinese round-back chair is the object, and in the design's interpretive process, Chinese furniture (PO-A) is the object, with the round-back chair as the carrier for interpretation. Through the designer's interpretation, the symbolic representation of the Chinese round-back chair's "heavenly roundness and earthly square" is manifested in the case chair, representing the design's interpretive process from object to interpretation to representation in the product symbol triangle. The Chinese seating is the designer's conceptual symbol, and the round-back chair is the communicative medium's symbolic carrier. Using the round-back chair as a representation, the object of the Chinese seating is reinterpreted within the new Oriental furniture brand's context.

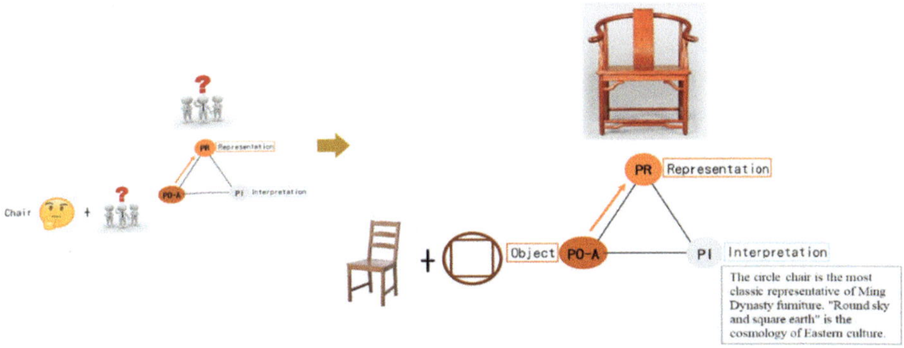

Fig. 2. Interpretation of the minimalist series chair in the product symbol triangle.

In short, Step 1 (Fig. 3): Translate the cultural codes of the research case's products to identify the source of the product's genes. Step 2 (Fig. 4): Then, through the aforementioned product symbol triangle analysis, define the object of each product and analyze the interpretative meanings of different products to organize the product representations of the research case brand. Step 3 (Fig. 5): Based on the different product representations, organize the relationship diagram of the paradigmatic and syntagmatic axes of each product. Step 4 (Fig. 6): Then, conduct continuous interpretive analysis using the product symbol triangle on the formed relationship diagram, making the symbolic features of the series products evident. This aids in ensuring greater consistency in the characteristics of new series products during design and development. Step 5 (Fig. 7): The rearrangement and combination of features within the same series products become more diverse, providing the brand with better attribute relationships when redeveloping products and preventing the misalignment of the genes between the products and the brand.

Through the translation of cultural code elements, Research Case products elucidate the genetic origins of the translated elements and then analyze the product's genetic origins using the product symbol triangle. This analysis aims to understand the object of the product and how cultural codes are interpreted in design to form symbol representations unique to Research Case products. The analyzed product representations are categorized using homogeneity sequence relationships and syntagmatic axis relationships of functional structures, leading to a paradigmatic axis-syntagmatic axis relationship diagram.

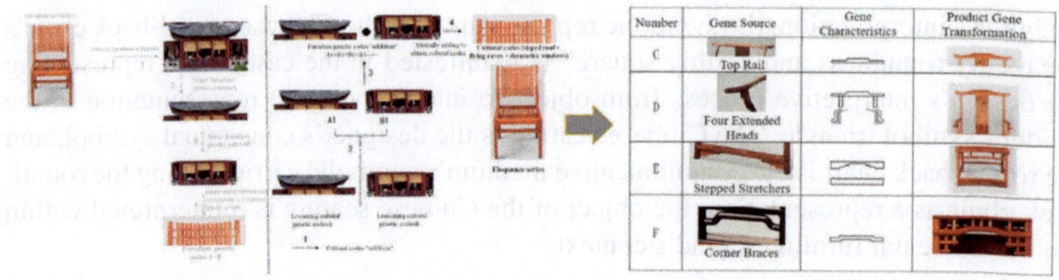

Fig. 3. Translate cultural codes to identify the source of product genes

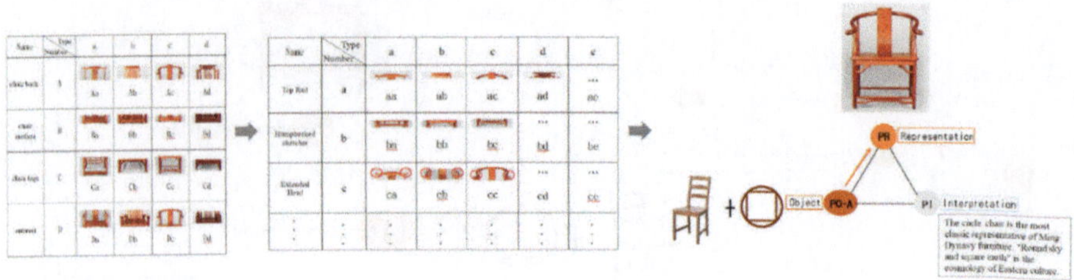

Fig. 4. Product symbol triangle analysis to define object, interpretation, representation

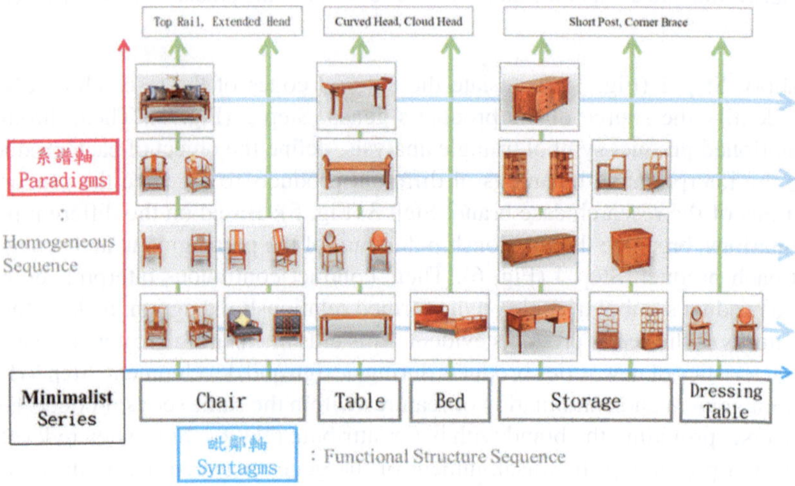

Fig. 5. Organize the relationship diagram of the paradigmatic and syntagmatic axes of products

Further clarification of categorized homogeneity and functional structure sequence products is conducted within the relationship diagram. Through continuous interpretive analysis of the product symbol triangle based on different selected objects, distinctive symbol features of various series of products are identified. These symbol features serve as references for product design and development applications, ensuring that developed products maintain the symbolic tone of different series and serve as essential references for product and brand tone consistency.

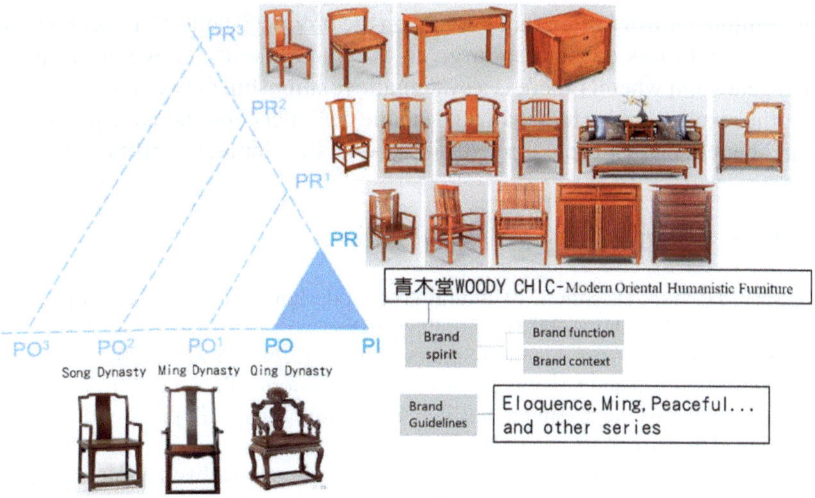

Fig. 6. The continuous interpretive analysis of the product symbol triangle.

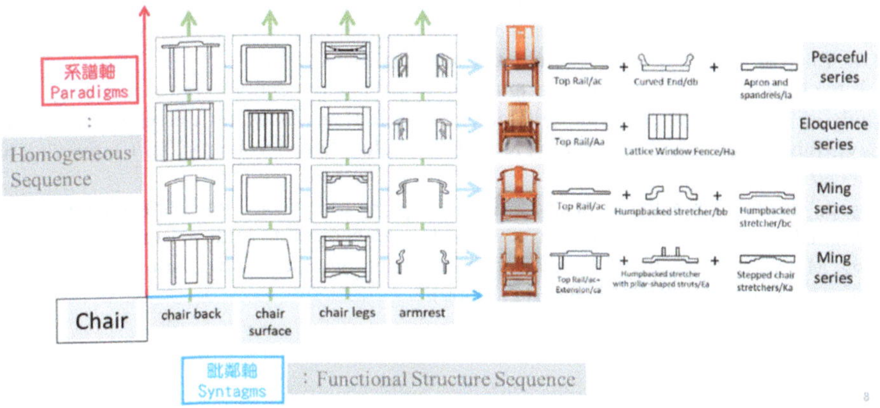

Fig. 7. The arrangement and combination of features within each series of products during design.

4 Conclusion

This study employs semiotic analysis to examine product characteristics and utilizes the product symbol triangle to reinterpret these characteristics into product representations. By incorporating the Research Case brand case study into the research process, it is found that Research Case often transforms product symbol features into those resembling traditional Chinese architecture and furniture, thereby conveying the brand spirit of "modern oriental furniture."

The study proposes innovative insights by combining the product symbol triangle with paradigmatic axis-syntagmatic axis analysis methods. It applies the continuous interpretive analysis method of the product symbol triangle to analyze the symbolic genetic relationships among products. Understanding that many companies in the wooden furniture industry often focus solely on the products themselves when launching new products, with less consideration given to the correlation between product genes and brand genes and whether there is a systematic relationship between new and existing products, the results of this study can also enable designers and industry practitioners to understand better and achieve consensus in product development and brand linkage.

References

Chai, C., Shen, D., Bao, D., Sun, L.: Cultural product design with the doctrine of the mean in confucian philosophy. Des. J. **21**(3), 371–393 (2018)

Fiske, J.: Introduction to Communication Studies. Routledge, New York (1990)

Hawkes, T.: Structuralism and Semiotics. University of California Press (1977)

Hawkes, T.: Structuralism and semiotics (Chen Yongkuan, Trans.). Taipei: South Book Series. (Original work published 1977) (1989)

Hsiao, M.-Y.: A study of Peirce's semiotic triangle in cultural creative product design (Ph.D. dissertation, Chaoyang University of Technology). National Digital Library of Theses and Dissertations in Taiwan (2014). https://hdl.handle.net/11296/gt2e6y

Jian, R.: Traditional Taiwanese furniture (Chen Shengzeng, Ed.). Zhongli City, Taoyuan County: Taoyuan County Cultural Heritage Association (2000)

Peirce, C.S.: The essential Peirce: Selected philosophical writings (1893–1913) (Vol. 2, N. Houser, J. R. Eller, & A. C. Lewis, Eds.). Indiana University Press (1998)

Wu, J.-K.: Research on establishing design patterns for brand identity in product styling: a case study of Taiwanese wood furniture [Doctoral dissertation, Chaoyang University of Technology]. Taiwan Thesis and Dissertation Knowledge Value-Added System (2023). https://hdl.handle.net/11296/7488qx

Zhang, J., et al.: Communication semiotics theory. Taipei: Yuan-Liou Publishing Co., Ltd. (Original work: Fiske, J. [1990]. Introduction to communication studies. Routledge) (2000)

Zhuang, M., Xiao, K.: A study on the formative elements and their manipulation in constructing the style of circle chairs. J. Design **1**(1) (2009)

Identification of Key Selection Criteria for Exoskeleton Applications in Automotive Production Through Analytic Hierarchy Process (AHP) Method

Woun Yoong Gan(✉), Raja Ariffin Bin Raja Ghazilla, Hwa Jen Yap, and Suman Selvarajoo

Centre for Sustainable and Smart Manufacturing, Universiti Malaya, Kuala Lumpur, Malaysia
wygan@siswa.um.edu.my

Abstract. Exoskeleton devices can augment human capabilities in daily activities through mechanically align with the human body, especially in the application of healthcare, industry and military. Exoskeleton technology currently remains uncommon among industry practitioners in Malaysia. This study focuses on addressing the gap by exploring the crucial criterial for exoskeleton implementation in automotive production through the Analytic Hierarchy Process (AHP) with experts from industry. Total there are 12 criteria was selected and after the evaluation by 10 experts from automotive industry through AHP method, the highest concern by industry was ease of maintenance and after sales support and the least is weight and size of exoskeleton. By highlighting key selection criteria, this research aims to enhance the suitability of exoskeleton characteristics in specific production tasks, moving beyond from the current focus on design different type of exoskeleton and develop a model for industry to select the suitable exoskeleton in the production tasks to increase the usage in the line.

Keywords: Exoskeleton technology · Automotive production · AHP · decision model

1 Introduction

1.1 Background

The exoskeleton device is a wearable device with purpose to help the users to perform various daily activities [1]. Now, exoskeletons are generally regarded as a technology that extends, complements, substitutes or enhances human function and capability or empowers the human limb where it is worn. Exoskeletons are commonly used in healthcare, industry and military applications [2], basically exoskeleton is works mechanically in parallel with human body and can be actuated passively and or actively [3]. However, exoskeleton technology is still considered new and fresh to the users in Malaysia and most of the industry practitioners are not familiar with the systems [4]. To implement more potential exoskeleton application in industry, there are several criteria that would be concern by industry to consider before applying the exoskeleton into the production.

Chin, C. M. M. et al. (Eds.): SSIEC 2024, PTT, pp. 15–19, 2025.
https://doi.org/10.1007/978-981-96-2806-3_3

Other than passive and active type of exoskeleton, exoskeleton can also divide into 3 categories which are full body, upper body and also lower body [5]. Full body is basically covering upper and lower body of human, while upper part of human body such as shoulder joint, elbow, wrist etc. [6] and lower body is covering waist down to play certain function such as supporting movement during walking. While there's several types of exoskeletons and complexity of tasks in production, often requiring multiple postures, makes it challenging for users to select the appropriate type of exoskeleton for specific production tasks.

Since the development of the Analytic Hierarchy Process (AHP), it is as a tool for the decision makers and researcher, and it is one of the most common multiple criteria decision-making tools [7]. The application of AHP in various domains like planning, dispute resolution, resource allocation, optimization and best alternative selection.This study is to highlight the criteria and rating through AHP to identify the important criteria of exoskeleton selection in the production so the usage of exoskeleton in automotive production will be increasing. Currently most of the exoskeleton research is focused on the design and application tasks of exoskeleton, however this research will be on the fundamental model for the selection of suitable exoskeleton in specific production tasks.

2 Methodology

This research was started with listing of selection criteria of exoskeleton. After the 12 selection criteria has been list out, the criteria were inserted to AHP method with expert from automotive production company participated in the research.

2.1 Industry Expert and AHP

The results were collected from 10 participants from the automotive industries that having working experience in automotive more than 5 years and having knowledge in engineering process or equipment. Five participants from passenger car production, four from bus production and one from Truck production. All the participants are from various departments like production, Research and Development (R&D), and procurement. The grouping of participants is as shown in Table 1. Participants provided their preference of exoskeleton criteria when considering applying exoskeleton into the production tasks. The feedback from participants was collected and rating was calculated through the geometric average and AHP method.

In this research, the feedback was collected by the participants through an online survey in Excel form. The participants were required to evaluate their options by comparing the 12 criteria to one another pairwise in terms of each criterion, with rating from point 1 to 9 with different weight of importance. These comparisons are made using a scale of relative importance derived from psychological judgments. From these pairwise comparisons, numerical values are calculated, which represent the priority of each element within the hierarchy. This process results in a set of priority scales that combine to give an overall ranking of the alternatives.

Table 1. Grouping of participants

Industry	Department	Participant
Passenger Car production	Procurement	1
	Production	3
Bus production	R&D	2
	Production	2
Truck production	Procurement	1
	Production	1
Total participant		10

3 Result and Discussion

The exoskeleton criteria are list as C1 to C12 as Table 2 and AHP rating from 1–9 and the geometry average from the respondents is listed in Table 3, while the ranking of the criteria is listed in Table 4 below:

Table 2. Exoskeleton criteria

Code	Exoskeleton Criteria
C1	Purchase cost
C2	Increase users' strength, endurance, and work quality
C3	Reduction of WMSD and fatigue
C4	User friendly, comfort and safety
C5	Durability and reliability of exoskeleton
C6	Degree of freedom of exoskeleton (user flexibility)
C7	Suitability for production requirement
C8	Weight and size of exoskeleton
C9	Sources of power and usage interval (battery life)
C10	Maintenance and operation cost
C11	Ease of maintenance and after sales support
C12	Training and competency requirement

From the calculation, the result of Consistency Ratio (C.R) is 0.049. The highest concerns for exoskeleton criteria from this study are Ease of Maintenance & After Sales Support, follow with Maintenance and Operation Cost, while the least concerning

Table 3. AHP model of exoskeleton selection criteria

No	C1	C2	C3	C4	C5	C6	C7	C8	C9	C10	C11	C12	w_i	w_i^0	Aw_i^0	λ_{mi}
C1	1.00	1.30	0.28	1.10	0.38	1.07	1.05	1.25	0.95	0.27	0.36	0.75	0.7051	0.0529	0.6762	12.7787
C2	0.77	1.00	1.16	0.66	0.66	2.13	0.59	0.96	1.15	0.32	0.30	2.22	0.8338	0.0626	0.8269	13.2154
C3	3.51	0.86	1.00	2.40	1.24	1.68	1.60	2.42	2.68	0.96	0.44	0.87	1.4024	0.1052	1.3780	13.0932
C4	0.89	1.52	0.42	1.00	1.12	2.40	2.20	2.12	1.30	1.41	0.40	1.52	1.1866	0.0890	1.1676	13.1116
C5	2.67	1.49	0.80	0.89	1.00	1.10	2.26	1.42	1.13	0.67	0.31	1.02	1.0776	0.0809	1.0204	12.6175
C6	0.93	0.47	0.59	0.42	0.69	1.00	0.54	1.61	0.94	0.52	0.35	1.11	0.6928	0.0520	0.6558	12.6127
C7	0.96	1.69	0.62	0.46	0.44	1.41	1.00	2.00	0.96	0.35	0.44	0.78	0.7971	0.0598	0.7620	12.7383
C8	0.80	1.02	0.41	0.47	0.71	0.64	0.50	1.00	1.70	0.40	0.23	1.26	0.6627	0.0497	0.6284	12.6361
C9	1.05	0.85	0.49	0.77	0.88	1.08	1.04	0.59	1.00	0.33	0.33	0.55	0.6913	0.0519	0.6443	12.4206
C10	3.69	3.07	1.05	0.71	1.48	1.97	2.87	2.49	3.00	1.00	1.62	3.68	1.9574	0.1469	1.9093	12.9979
C11	2.76	3.36	2.33	2.49	3.21	2.93	2.26	4.31	4.21	0.62	1.00	5.16	2.5195	0.1891	2.3907	12.6437
C12	1.34	0.45	1.14	0.67	0.98	0.90	1.28	0.99	2.49	0.27	0.19	1.00	0.7990	0.0600	0.7966	13.2852

Table 4. Ranking for criteria

Criteria	Ranking
Ease of maintenance and after sales support	1
Maintenance and operation cost	2
Reduction of WMSD and fatigue	3
User friendly, comfort and safety	4
Durability and reliability of exoskeleton	5
Increase users' strength, endurance, and work quality	6
Training and competency requirement	7
Suitability for production requirement	8
Purchase cost	9
Degree of freedom of exoskeleton (user flexibility)	10
Sources of power and usage interval (battery life)	11
Weight and size of exoskeleton	12

criteria is Degree of freedom of exoskeleton (user flexibility), sources of power and usage interval (battery life) and Weight & Size of exoskeleton.

Some respondents expressed their concern regarding the maintenance and after-sales support of exoskeletons, attributing this to worries about the stability and reliability of the new technology in production. Therefore, support is deemed crucial during downtimes and for maintenance needs. Additionally, some respondents also highlighted that, compared to the initial purchase costs of exoskeletons, the outcomes and benefits offer by exoskeleton to the workers are of greater concern when considering their application into the production tasks.

4 Conclusion

As conclusion, this research found that the top three criteria for exoskeleton selection are Ease of Maintenance & After Sales Support, followed by Maintenance & Operation Cost, and Reduction of WMSD & Fatigue. It somehow shows that the industry is concerned on the long- term sustainable usage of exoskeleton and the ability to reduce work related injuries in the production. By utilizing the AHP with input from different experts from industry, this research has successfully highlighted crucial criteria for selecting suitable exoskeleton in the suitable tasks in the production lines. This approach major to addressed the existing gap in usage and familiarity with exoskeleton technology. The establishment of a decision-making model based on these findings could significantly increase the adoption and effectiveness of exoskeletons in enhancing human capabilities within the industry.

5 Future Development

The outcome of this findings will be use as a fundamental for exoskeleton selection decision model as to select suitable exoskeleton in complex production tasks.

Acknowledgement. The author would like to thank the Centre for Sustainable and Smart Manufacturing, Universiti Malaya, Kuala Lumpur, Malaysia for the support of the research and industry experts for their valuable feedback.

References

1. Antwi-Afari, M., et al.: Assessment of a passive exoskeleton system on spinal biomechanics and subjective responses during manual repetitive handling tasks among construction workers. Saf. Sci. **142**, 105382 (2021)
2. Proud, J.K., et al.: Exoskeleton application to military manual handling tasks. Hum. Factors **64**(3), 527–554 (2022)
3. Anam, K., Al-Jumaily, A.: Active exoskeleton control systems: state of the art. Procedia Eng. **41**, 988–994 (2012)
4. Gan, W.Y., et al.: Industrial practitioner's perception on the application of exoskeleton system in automotive assembly industries: a Malaysian case study. Heliyon **10**(4), e26183 (2024)
5. Pamungkas, D., et al.: Overview: types of lower limb exoskeletons. Electronics **8**, 1283 (2019)
6. Olar, M.-L., Leba, M., Risteiu, M.: Exoskeleton - wearable devices. Lit. Rev. MATEC Web Conf. **342**, 05005 (2021)
7. Vaidya, O., Kumar, S.: Analytic hierarchy process: an overview of applications. Eur. J. Oper. Res. **169**, 1–29 (2006)

Improving Security Enhancement for Cell Phone Using Finger Vein Biometrics System

Kayode A. Akintoye(✉) and Sunday Akinwamide

Department of Computer Science, The Federal Polytechnic, Ado-Ekiti, Nigeria
akintoye_ka@fedpolyado.edu.ng

Abstract. The swift progress of technology in enhancing the security of mobile devices has resulted in the extensive utilization of personal identification numbers (PINs), passwords, and gridlock patterns to safeguard phones and associated social applications. Nevertheless, these conventional security measures are riddled with susceptibilities, such as password pilferage, memory lapses in authentication details, and user mistakes in grid pattern creation. To tackle these problems, a range of biometric methods, such as fingerprint and facial recognition, have been put into practice. Notwithstanding these progressions, there are nonetheless notable deficiencies in security. Due to the rising occurrence of cell phone incursions and theft, there is an urgent requirement for a strong security system that not only defends data but also shields the device itself. This study presents a resilient finger vein biometric security system as a more effective substitute for knowledge-based and password-based authentication techniques. The method employs infrared (IR) light transmission to capture the vein patterns and shadows resulting from the different thicknesses of finger muscles, bones, and tissues. Research on human finger vein biometrics suggests that every person has a distinct vein pattern, which enhances the level of security. The simulations we conducted indicate an identification accuracy rate of 93.82%, indicating that this technique provides a substantial enhancement in preventing unauthorized access and theft of cell phones.

Keywords: Cell Phone Security · Finger Vein Biometrics · Biometric Authentication

1 Introduction

As cell phones devices have become integral to daily life, they increasingly serve as repositories for sensitive personal and corporate information. This has amplified the need for advanced security measures to protect against unauthorized access and data breaches. Traditional authentication methods, such as personal identification numbers (PINs), passwords, and pattern locks, remain prevalent but have shown significant weaknesses. Studies have demonstrated that these methods are vulnerable to various attacks, including shoulder surfing, brute-force, and social engineering attacks, rendering them insufficient in safeguarding mobile devices (Smith, 2020; Johnson & Wang, 2019).

Chin, C. M. M. et al. (Eds.): SSIEC 2024, PTT, pp. 20–26, 2025.
https://doi.org/10.1007/978-981-96-2806-3_4

To overcome the limitations of knowledge-based authentication methods, biometric systems have emerged as a viable alternative, leveraging unique physical or behavioral traits for identity verification. Common biometric systems include fingerprint recognition, facial recognition, and iris scanning. Although these systems provide enhanced security compared to traditional methods, they are not without flaws. For example, fingerprint sensors can be deceived by synthetic fingerprints, and facial recognition can be circumvented using high-quality photographs or masks (Lee et al., 2021; Zhang & Liu, 2018). Additionally, these systems often suffer from performance issues in non-ideal conditions, such as poor lighting or dirty sensors (Fig. 1).

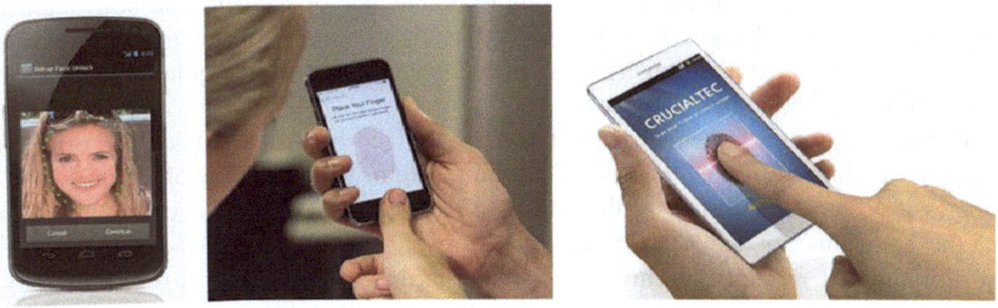

Fig. 1. Biometrics in smartphones: (a) Face unlock method of Android, (b) Touch ID of iOS, and (c) Biometric TrackPad of CrucialTec

Recent studies have highlighted the potential of finger vein recognition technology in various applications, including banking, access control, and now, mobile device security (Chen et al., 2020; Nakamura et al., 2018). With an accuracy rate of over 93% in controlled environments, finger vein biometrics is well-positioned to address the growing concerns surrounding cell phones device security (Akintoye et al., 2018, Wang et al., 2019).

This paper explores the integration of finger vein biometric systems into cell phones devices as means to enhance security. By evaluating the underlying technology, potential implementation strategies, and the security benefits it offers, this study aims to demonstrate how finger vein biometrics can serve as a robust alternative to existing authentication methods. The findings suggest that this innovative approach could significantly improve the protection of cell phones devices, reducing the risk of unauthorized access and data breaches.

2 Literature Review

The rapid evolution of mobile technology has led to increased concerns about security, particularly as mobile devices are now used for a wide range of sensitive activities, including banking, communication, and personal data storage. This literature review explores the limitations of traditional and current biometric authentication methods and highlights the potential of finger vein biometrics as a robust solution for enhancing cell phone security.

Traditional authentication methods, such as passwords, PINs, and pattern locks, remain prevalent despite their well-documented vulnerabilities. These methods are susceptible to various forms of attack, including phishing, brute-force attacks, and social engineering. According to research by Bonneau et al. (2012), passwords are often weak due to poor user practices, such as reusing passwords across multiple sites or choosing easily guessable passwords. These weaknesses have led to an increased interest in more secure forms of authentication.

Biometric authentication systems have gained traction as a more secure alternative to traditional methods, leveraging unique physiological and behavioral characteristics of individuals. Fingerprint recognition has become one of the most widely adopted biometric technologies in smartphones due to its balance of security and convenience (Maltoni et al., 2009). However, as shown by Matsumoto et al. (2002), fingerprint sensors can be deceived by artificial fingerprints created using materials like gelatin or silicone, raising concerns about their robustness.

Facial recognition systems, particularly popularized by Apple's Face ID, offer an alternative to fingerprints, using sophisticated algorithms to analyze facial features. Nevertheless, recent studies, such as the one by Nguyen et al. (2021), have demonstrated that these systems can be by-passed using high-resolution images or 3D-printed masks, questioning their reliability in high-security scenarios. Iris recognition is another biometric method known for its high accuracy. However, it is less commonly used due to its reliance on specialized hardware and susceptibility to environmental factors like lighting, which can affect accuracy (Daugman, 2004).

Finger vein recognition technology represents a promising advancement in biometric security. The method uses near-infrared (NIR) light to scan the unique vein patterns within a user's finger, providing a highly secure means of authentication that is difficult to replicate or spoof (Kono et al., 2002). This internal biometric feature is not visible to the naked eye and is protected by the skin, making it inherently more secure than external biometrics like fingerprints or facial features.

Recent studies have shown the effectiveness of finger vein biometrics in various applications. For instance, Miura et al., 2007 demonstrated the high accuracy of finger vein recognition, with a false acceptance rate (FAR) as low as 0.0001% and a false rejection rate (FRR) of 0.01%, significantly outperforming other biometric systems. Additionally, Lee et al. (2017) explored the potential for integrating finger vein technology into mobile devices, highlighting the feasibility of miniaturizing the necessary hardware without compromising performance.

Despite its advantages, the adoption of finger vein biometrics in mobile devices faces challenges. The cost of integrating near-infrared sensors and cameras into smartphones is higher compared to existing biometric solutions, which could limit its initial adoption (Zhao et al., 2019). Furthermore, user acceptance is a critical factor, as consumers may be hesitant to adopt a new biometric system that they are unfamiliar with or that may require adjustments to the device's design.

Comparative studies underscore the potential of finger vein biometrics to outperform existing biometric systems in terms of security and resistance to spoofing. A study by Wang et al. (2019) compared finger vein recognition with fingerprint and facial recognition systems, concluding that finger vein biometrics provided superior protection against a wide range of attacks. The study also noted the additional layer of security provided by the internal nature of vein patterns, which are not easily accessible or visible to attackers.

Moreover, finger vein recognition offers a unique combination of security and user convenience. Unlike fingerprint recognition, which can be affected by skin conditions, or facial recognition, which can be impacted by lighting, finger vein recognition is less prone to environmental factors, making it a reliable option for everyday use (Yang et al., 2020). This robustness, combined with its high level of security, positions finger vein biometrics as a leading candidate for enhancing mobile device security.

3 Methodology

The objective of this study is to develop and evaluate a finger vein biometric authentication system tailored for mobile devices. This section outlines the methodological approach taken to design, implement, and test the system, including the selection of hardware, data collection, system architecture, and evaluation metrics.

3.1 System Design and Hardware Selection

The finger vein biometric system was designed to be integrated into mobile devices, requiring careful selection of hardware components that balance accuracy, cost, and miniaturization. The system employs a near-infrared (NIR) imaging sensor to capture finger vein patterns, as NIR light penetrates the skin and reveals the underlying vascular structure, which is unique to each individual.

3.2 Data Collection and Preprocessing

To develop and validate the finger vein recognition system, a dataset of finger vein images was collected from a diverse group of participants. The dataset included images from 100 participants, with multiple images taken from each finger under varying conditions to account for natural variations in finger positioning and lighting.

Participants were selected to represent a wide demographic range, including variations in age, gender, and skin tone, to ensure the system's robustness across different populations. Images were captured in a controlled environment with consistent lighting and background conditions to minimize external noise. Each participant's finger was imaged multiple times, and slight variations in finger positioning were encouraged to simulate real-world usage scenarios.

Preprocessing steps included image enhancement techniques to improve the contrast between the vein patterns and the surrounding tissue. Techniques such as histogram equalization and contrast stretching were employed to achieve this. Images were normalized to a standard size and orientation to ensure consistency across the dataset. This involved aligning the images based on key reference points in the finger structure.

The next step in the methodology involved extracting features from the preprocessed finger vein images and using these features for pattern recognition. To enhance the security of finger vein identification on cell phones, the study proposes a novel approach that combines hierarchical centroid features with statistical pixel features. The experimental method, dubbed Combined Feature Extraction Method (CFM), integrates two techniques: Hierarchical Centroid Features Extraction Method (HCM) and Pixel Distribution-based Feature Extraction Method (PDM). By fusing multiple sources of information during feature extraction, the study aim to boost identification performance (Kolivand et al., 2023). The CFM is applied to evaluate the identification scheme's performance, generating a unique set of features from each image in the dataset for the classifier to make accurate decisions. Figure 2 illustrates the research design's block chart for feature extraction.

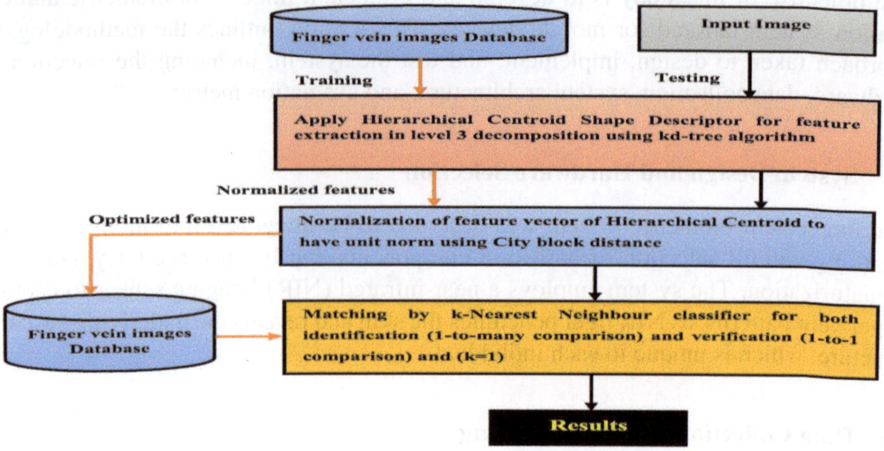

Fig. 2. Research design for feature extraction

4 Discussion

The final phase involved integrating the finger vein biometric system into a prototype mobile device and conducting a series of tests to evaluate its performance. The section first presents the characteristics of the finger vein database (FV-USM) using the Combined Feature Extraction Method (CFM). Specifically, feature vectors from the index, middle, and ring fingers of both left and right hands are utilized for enrollment purposes. To assess the accuracy of the finger vein identification method, we conduct two types of matching tests: genuine and imposter. The genuine matching test measures the pairwise matching distance between images from the same class, while the imposter matching test calculates the distance between image pairs from different classes. These tests evaluate the effectiveness of the identification method in distinguishing between authentic and non-authentic matches (Table 1).

Table 1. Identification and EER results for CFM feature using FV-USM database

Finger Type	Identification Rate	Verification Rate (%)			Mean Time
1st Session	(%)	FAR	FRR	EER	(seconds)
1	81.3008	0.1532	9.6992	3.4262	0.0639
2	85.3659	0.1199	7.6341	2.3770	0.0634
3	82.9268	0.1399	8.0732	3.6066	0.0636
4	93.8224	0.0799	4.7561	2.9180	0.0630
All (1 – 4)	81.9106	0.0368	9.0894	3.0631	0.0785

The experiments conducted to determine the most suitable distance metric for identification and verification experiments in this paper were conducted. The study evaluated three distance metrics: Euclidean distance, City block distance, and Cosine distance. The results presented in Table 2, show an average of five trials for each distance metric. Based on these results, the study selected the City block distance as the optimal choice for use with the nearest neighbor classifier in this research, due to its superior performance.

Table 2. The average of 5 run for different distance types

Distance	Identification Rate (%)	Verification Rate (EER %)
Cosine	90.50	2.11
Euclidean	91.88	1.83
City block	93.82	1.11

5 Conclusion

This paper has presented a comprehensive explanation of the enhancement of cell phone security through finger vein biometric authentication systems. The proposed method improves image quality for finger vein feature extraction, leading to enhanced human identification performance. The feature extraction method is straightforward and computationally efficient. The study adopted City block distance for the k-nearest neighbor (KNN) classifier due to its superior accuracy compared to other distance metrics. Our evaluation metrics showed a significant improvement in identification accuracy, with a high accuracy rate of 93.82% and an equal error rate (EER) of 1.11% for verification. These results were achieved through testing on publicly available finger vein databases, demonstrating the effectiveness of the proposed method.

References

Akintoye, K.A., Rahim, M.S.M., Abdullah, A.H.: Challenges of finger vein recognition system: a theoretical perspective. Int. J. Emerg. Technol. Adv. Eng. **8**(2), 196–204 (2018)

Bonneau, J., Herley, C., Van Oorschot, P.C., Stajano, F.: The quest to replace passwords: a framework for comparative evaluation of web authentication schemes. In: IEEE Symposium on Security and Privacy, pp. 553–567 (2012)

Cheng, X., Wang, P., She, C.: Biometric identification method for heart sound based on multimodal multiscale dispersion entropy. Entropy **22**, 1–21 (2020)

Daugman, J.: How iris recognition works. IEEE Trans. Circ. Syst. Video Technol. **14**(1), 21–30 (2004)

Johnson, K., Wang, L.: The vulnerabilities of knowledge-based authentication in the age of biometrics. Cybersecurity Rev. **8**(2), 89–101 (2019)

Kolivand, H., Akintoye, K.A., Asadianfam, S., Rahim, M.S.: Improved methods for finger vein identification using composite median-wiener filter and hierarchical centroid features extraction. Multim. Tools Appl. **82**(21), 31913–31944 (2023)

Kono, M., Ueki, H., Umemura, S.-I.: Near-infrared finger vein patterns for personal identification. Appl. Optics. **41**(35), 7429–7436 (2002)

Lee, H., Kim, S., Park, J.: Biometric authentication systems: strengths and weaknesses. Int. J. (2021)

Lee, J., Kim, E., Yang, K.: A study on miniaturized finger vein recognition sensor for mobile devices. In: Proceedings of the 8th International Conference on Awareness Science and Technology (iCAST), pp. 430–434 (2017)

Maltoni, D., Maio, D., Jain, A.K., Prabhakar, S.: Handbook of Fingerprint Recognition. Springer Science & Business Media (2009)

Matsumoto, T., Matsumoto, H., Yamada, K., Hoshino, S.: Impact of artificial "gummy" fingers on fingerprint systems. In: Proceedings of SPIE, Vol. 4677, pp. 275–289 (2002)

Miura, N., Nagasaka, A., Miyatake, T.: Extraction of finger-vein patterns using maximum curvature points in image profiles. IEICE Trans. Inf. Syst. **90**(8), 1185–1194 (2007)

Nakamura, T., Goverdovsky, V., Mandic, D.: In-ear EEG biometrics for feasible and readily collectable real-world person authentication. IEEE Trans. Inf. Forensics Secur. **13**, 648–661 (2018)

Nguyen, K., Yadav, P., Halder, A., Ramanathan, M.: Face recognition with 3D printed masks. IEEE Trans. Inf. Forensics Secur. **16**, 3757–3771 (2021)

Smith, J.: Mobile security: an overview of current authentication methods. J. Inf. Secur. **15**(3), 123–145 (2020)

Wang, L., Le, T., Pham, D.: Comparative analysis of biometric technologies for mobile security: a study of finger vein, fingerprint, and facial recognition. Comput. Secur. **85**, 113–126 (2019)

Yang, Y., Zhang, Z., Wang, L. Environmental robustness of finger vein biometric systems: a comprehensive analysis. J. Inf. Secur. **11**(2), 87–99 (2020)

Zhao, Y., Zhang, W., Wang, R.: Challenges in integrating finger vein recognition technology into mobile devices. J. Mobile Comput. **8**(3), 123–135 (2019)

Community Local Resource Integration and Agricultural Waste Application Design

Yann-Long Lee[1](✉), Yu-Ching Hung[2], Chuan-Po Wang[3], and Chun-Kuan Wu[1]

[1] Department of Industrial Design, Chaoyang University of Technology, Taichung City, Taiwan
yannlee@cyut.edu.tw

[2] Faculty of Industrial and Manufacturing Technology and Engineering (FTKIP), Universiti Teknikal Malaysia Melaka (UTeM), Durian Tunggal, Malaysia

[3] Department of Creative Product Design, National Taichung University of Science and Technology, Taichung 404348, Taiwan

Abstract. Agricultural waste is inedible parts or by-products during agricultural production or processing. These wastes cause environmental impact and are a headache for local governments. This study mainly focuses on grapevine agricultural waste to see whether grapevine waste can be converted into useful commodities through the integration of local resources to achieve the goal of the circular economy. This study uses on-site visits, participatory design, and industrial design methods to reuse and design grapevine agricultural waste that originally seemed useless. Combined with local resources, we integrate communities, returning youth, temples, manufacturers, etc., to develop a series of daily necessities. Agricultural waste has once again become an indispensable product in life. As a result of the research, a series of biochar products, tableware, religious items, etc., have been developed and won many awards. Develop a feasible production model for the circular economy and serve as a model for sustainable product design.

Keywords: Agricultural Waste · Product Design · Agricultural Waste Reuse · Biochar Products · University Social Responsibility (USR)

1 Introduction

In recent years, human activities have greatly impacted the natural environment system, resulting in many impacts on food, water resources, land, etc. related to climate change. Therefore, the United Nations adopted the 2030 Sustainable Development Agenda in 2015 and proposed 17 core goals (Sustainable Development Goals, SDGs) for the world to move towards sustainable development. More than half of them are related to agriculture (Cai WT, 2022).

At the same time, Taiwan's Ministry of Environmental Protection, Agriculture and Economic Affairs has also formulated relevant agricultural waste reuse regulations in response to this international trend and in compliance with the principle of sustainable utilization of natural resources. Promote industry players to adopt new agricultural production methods such as bioeconomy, circular economy, organic agriculture, and smart agriculture. And then actively face the specific actions to achieve the 2050 net-zero carbon emission target under the post-Kyoto Protocol.

Chin, C. M. M. et al. (Eds.): SSIEC 2024, PTT, pp. 27–33, 2025.
https://doi.org/10.1007/978-981-96-2806-3_5

1.1 Agricultural Waste

Among these actions, agricultural waste reuse is a very hot topic. Agricultural waste refers to the waste generated during agricultural production activities and agricultural product processing. It can be divided into agricultural waste, fishery waste, livestock waste, wholesale market waste, and food processing waste.

Agricultural waste includes cereals, crops, fruits and vegetables, mushroom cultivation, bark, and other waste. Agricultural material waste includes materials such as fertilizer bags, seedling bags, waste pesticide containers, PE films, PVC films, and shade nets (Chen RX, 2018). Most of these wastes are non-perishable and cannot be used as raw materials for organic fertilizers. Farmers used these wastes as feed or applied to agricultural land in the past. However, because the wastes have high moisture content and are perishable, they must be treated appropriately to avoid environmental pollution.

Agriculture produces a variety of wastes during the production process, which is divided into two categories according to their properties: (Huang YZ; Chen PT and Yu QW, 2016).

Decomposability
Most decomposable wastes, such as rice straws, fallen fruits, branches, etc., can be left in the field for natural decomposition, returned to nature, or reused as fertilizers, feed ingredients, etc.

Non-decomposable
Plastic materials, such as films, ropes, and nets, are the most common non- degradable waste. To avoid polluting the environment, the above-mentioned items should be properly collected and handed over to qualified disposal units after being discarded. Do not discard them randomly or burn them in the open. Those who violate the regulations will be fined according to the law.

Agriculture in Taiwan produces approximately 4.6 to 5.2 million tons of agricultural waste a year (Li GJ, 2022). The waste generated is not only difficult to handle, but it can also cause harm to the environment if not handled properly. Therefore, how to properly utilize it has become an emerging green business opportunity today.

Currently, the most common way to use agricultural waste in Taiwan is to chop, stir, and ferment these agricultural waste raw materials into organic compost and use it for recycling on crops.

1.2 Survey on Commercially Available Agricultural Waste Reuse Products

Circular Economy (CE) is a production and consumption model that pursues sustainable shared use, reuse, remake, recycling, and reuse. Let products be recycled through redesign, better business models, and efficiency improvements. The theory of circular economy has been proposed by experts from various countries as early as the 1970s, including concepts such as “regenerative design,” “cradle to cradle,” and “sustainable culture,” with the aim of reducing resource waste and achieving complete recycling (Wu PZ, 2023).

There are already quite a few agricultural waste reuse products on the market, and they are becoming more and more prosperous. The following are some of the results of this research investigation (Table 1).

Among them, many communities and enterprises take the "circular economy" as the starting point, regenerating the waste left behind by local agricultural specialties and developing them into local brands through the introduction of technology and design. It solves the environmental pollution problem caused by burning large amounts of waste and brings commercial and tourism income while considering sustainable development. Because the success of the brand has made places known and known to more people, similar local cases are continuing to increase (Zheng CY & Chen FR, 2021).

Table 1. Survey form on commercially available agricultural waste recycling products

Agricultural Waste Products	Description
Allbirds - Wool Runner	It uses merino wool, natural eucalyptus fiber, and sugarcane waste as raw materials and can be decomposed naturally
Adidas- Stan Smith Mylo	Mylo mushroom leather can be grown from mushroom mycelium in just 2 weeks. The feel and texture are similar to real leather. This material is not yet completely biodegradable
Spanish sneaker brand "Saye"	Leather made from mango and other fruits, and sneaker linings made from bamboo fiber
Pineapple Leaf Vegetarian Wallet "OhGio"	Made of pineapple leaf fiber, 100 kg of leaves can only produce 1 kg of fiber, and then the embroidery is cleverly combined on the Phoenix fiber wallet. Use pineapple fiber to replace animal skin, which is fashionable and does not kill animals
Rush Remade "Edible Chopsticks - Tatami"	Kumamoto Prefecture, Japan uses 100% edible rushes to make "edible chopsticks-tatami". Each chopstick is handmade by craftsmen, who knead it into dough and bake it slowly at low temperature. It has high nutritional value and dietary fiber. The content is about 60 times that of lettuce. After using chopsticks, it can be eaten as if it is a biscuit
Mango peel and fruit bagging turned into recycled paper	In order to promote circular agriculture, a community in Tainan, Taiwan, uses the bags and mango peels of abundant mango, star fruit, and other fruits to make recycled paper, recycled dishes, and other products by hand-making paper

(*continued*)

Table 1. (*continued*)

Agricultural Waste Products	Description
"Hypnose" tableware	Hypnose tableware products are made entirely from waste materials. Waste soil produced by the manufacturing of colored porcelain tableware is collected. New materials with uniform textures are created through a simple vacuum pressing process. Each tableware bowl is unique and cannot be copied
"Re-ing" bamboo fiber plate	Re-ing uses patented plastic-free Taiwanese bamboo fiber to produce the plates and designs them with finishing touches. The plates are colored with natural food pigments, and the uneven surface enhances structural stability, facilitates stacking, and creates a dazzling reflective visual effect

1.3 University Social Responsibility (USR)

Taiwan's Ministry of Education implements the "University Social Responsibility (USR) Project". Focusing on local connections, talent cultivation, international connections, and other aspects and various issues, it plays an important role and promoter of social participation. Encourage the use of professional knowledge and creativity to improve the learning-application gap; promote local identity and development, and then move towards the vision of international integration (USR Promotion Center, 2024).

The concept of USR has been popular internationally for a long time. The United States and Canada began to promote universities' responsibility for sustainable development around 2010. The European Union promoted the European Union University Social Responsibility (EU-USR) between 2012 and 2014 to promote USR (Feng SE, 2023).

The goal of USR is to enable universities to pay more attention to local issues, provide professional knowledge and technology, and make more contributions to regional development. Through USR, the university will cooperate with local society and industry to jointly explore problems and design solutions so that the university's research results can directly feedback to society, thereby promoting regional development.

2 Methods

This study uses on-site visits, participatory design, and industrial design methods to reuse and design grapevine agricultural waste that originally seemed useless. A series of daily necessities are developed by combining local resources to integrate communities, returning youth, temples, manufacturers, etc. Agricultural waste has once again become an indispensable product in life (Table 2).

Table 2. Research and survey methods

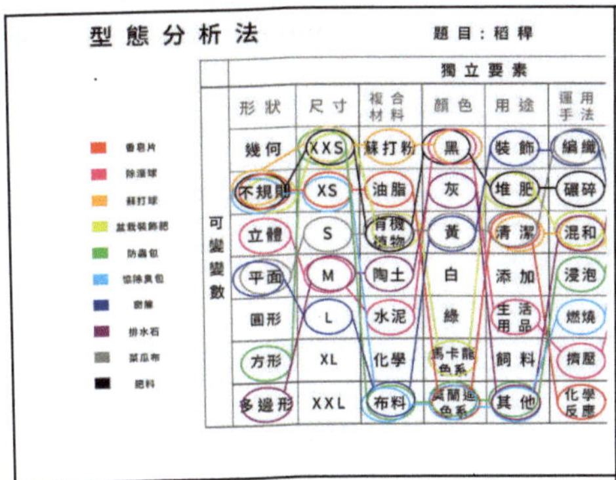

Pattern Analysis Method

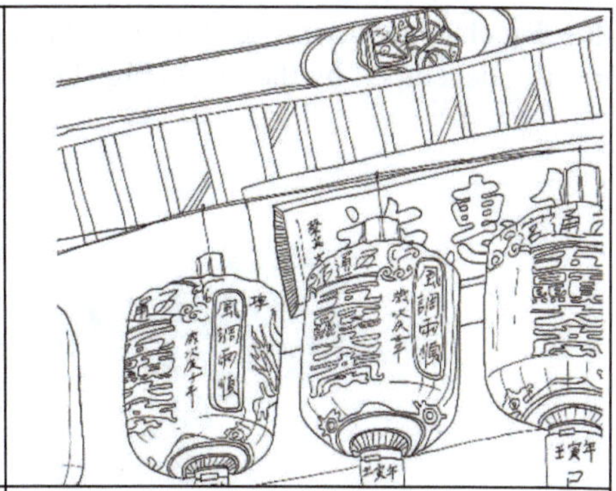

Field Visit Method

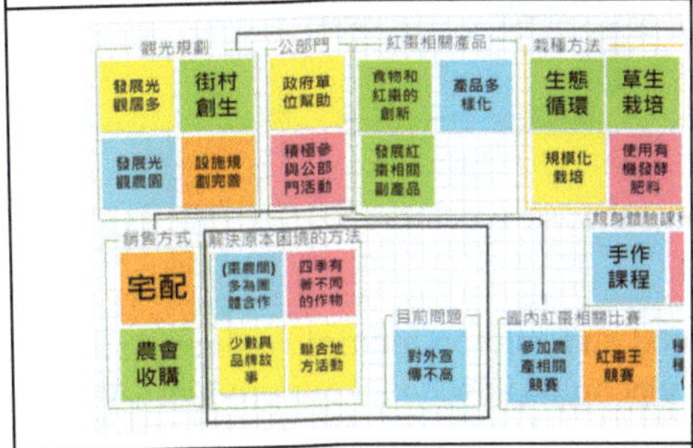

KJ Method

Community Survey Comparative Method

3 Results and Discussion

Changhua is located in the middle of western Taiwan, this place is rich in grapes. More than half of the local residents make a living by growing grapes. After the grapes are harvested, there will be a lot of grapevines. Every year, a piece of land will result in about 250 metric tons of agricultural waste in the Xing-Xing community. Many grape planting areas in Dacun Township produce about 14,000 metric tons of grapes and nearly 1,500 metric tons of grapevines. At present, 95% of the treatment methods are open-air burning. Farmers are faced with such a huge amount of agricultural waste, and how to deal with it is a big problem.

According to the book Cradle to Cradle, the researchers believe there is no natural waste. Therefore, observe daily life to find out more possibilities of biochar and integrate biochar into our daily lives so that agricultural waste will enter our lives. Therefore, the researchers integrate biochar into industries at all levels. First, use the characteristics of biochar to make odor-absorbing bags. Next, biochar can be reconstructed into other products, such as incense sticks, potted plants, soap, etc., so that biochar can enter our daily lives and be used by the public.

Finally, biochar is applied to the powder metallurgy process to make necklaces and collar buttons, allowing biochar to enter the high-end market and create higher added value. Biochar has also passed heavy metal testing. These are featured products of biochar currently developed by the team (Fig. 1).

Fig. 1. Various agricultural waste products

A total of 20 products have been developed and sold on multiple online platforms. We regularly visit different markets to let more people know about our products. We also go into the community to learn and grow together with the elders in the community. At the same time, it also injects youthful vitality into the community, builds local sentiment, and gains local identity.

Local residents and consumers also recognize our products. During this period, the team also accepted reports and interviews from large and small news media and established FB fan pages, IG, self-media, and online channels. Let more people understand the problem of agricultural waste so that everyone can solve it together and contribute to a sustainable environment.

The "Technological Discovery at the Foot of Bagua Mountain" USR project integrates the design energy, technology improvement, and marketing guidance of Chaoyang University of Science and Technology. Construct a design platform for agricultural waste reuse, biochar application technology guidance, brand building, and marketing innovation in Dacun Township, Changhua County. It also links regional industries, integrates the curriculum and the youthful vitality of students, constructs a teaching field with local characteristics, promotes sustainable design in applying scientific grapevine biochar, and has achieved quite fruitful results.

Subsequently, students have been trained to establish the company "Many Vines," which can continue to operate the fruitful results of USR. We can use the project results as a basis to develop and expand, discover more problems that need assistance, and help community industries more completely. During this period, students studied in various ways, developed various products, won awards at invention exhibitions, reported in the news, etc. They won numerous awards and were recognized by all walks of life.

4 Conclusion

The linear economy can be said to be a waste of resources. Production activities continue to consume resources, resulting in excessive manufacturing and rapid depletion of global environmental resources. The circular economy simulates the characteristics of the food chain of the earth's ecology, creating an endless cycle of production, consumption, recycling, and reuse. We can fundamentally reduce raw materials consumption, improve raw materials' economic benefits, and move towards an ecological system with endless cycles (Huang SW, 2017).

As a result of the research, a series of biochar products, tableware, religious items, etc., have been developed and won many awards. Develop a feasible production model for the circular economy and serve as a model for sustainable product design.

Acknowledgements. Special thanks to EASON Quantum Fiber Co., Ltd., WU TONG GONG Temple, Xing-Xing Community Development Association, Taiwan Ministry of Education USR Program, and Agency of Rural Development & Soil and Water Conservation, MOA for their assistance.

References

Cai, W.-T.: Recycling of agricultural waste, Department of Biological Resources, College of Agriculture, National Pingtung University of Science and Technology, (in Chinese) (2022). https://agriculture.npust.edu.tw/2022/11/agr_waste/. Accessed 20 Mar 2024

Chen, R.-X.: Development of organic waste reuse and rapid treatment technology. Sci. Mon. Iss. **577** (2018). (in Chinese)

Feng, S.-E.: What is USR University Social Responsibility? Why should all college students understand? ESG vision webpage (2023). https://esg.gvm.com.tw/article/25878. Accessed 1 Mar 2024

Huang, S.-W.: Opportunities and challenges in promoting circular economy in Taiwan. Taiwan Econ. Res. Monthly **40**(6) (2017). (in Chinese)

Huang, Y.-Z., Chen, P.-T., Yu, Q.-W.: Discussing my country's circular agriculture development strategy from the perspective of international circular economy trends. Agric. Biotechnol. Ind. Q. **46** (2016). (in Chinese)

Li, G.-J.: Promoting circular agriculture innovation and technology assistance. Agr. Politics Agr. Situation **111**, 355–366 (2022). (in Chinese)

USR Promotion Center: Introduction to University Social Responsibility Programs (2024). https://usr.moe.gov.tw/tw/about/usr. Accessed 1 Mar 2024

Pei-Zhen, W.: Turning nature into magic - circular economy brings new solutions to local areas. Taiwan Econ. Res. Mon. **46**(3), 110–117 (2023). (in Chinese)

Zheng, C.-Y., Chen, F.-R..: Research on the local brand strategy of agricultural waste regeneration transformation. In: 2021 ICWVCD International Symposium on Visual Communication Design and Basic Shaping Forum (2021). (in Chinese)

Achieving Corporate Sustainable Development: Leveraging on Process Management Practices and Knowledge Management Capabilities

Charis Samuel Solomon Koipillai[1], Marini Nurbanum Mohamad[2](✉), and Ramayah Thurasamy[3]

[1] Faculty of Industrial Management, Universiti Malaysia Pahang Al-Sultan Abdullah, Pekan, Pahang, Malaysia
charis@ump.edu.my
[2] Faculty of Business and Economics, Universiti Malaya, Kuala Lumpur, Malaysia
marininur@um.edu.my
[3] School of Management, Universiti Sains Malaya, Gelugor, Pulau Pinang, Malaysia
ramayah@usm.my

Abstract. Many businesses have undertaken various sustainable development measures to moderate the adverse environmental and social effects of their operations. This paper aims to identify and provide evidence pertaining to the relationship of process management practices (process design, process improvement and process control) towards corporate sustainable development (environmental development, economic development and social development) and the moderating effect of knowledge management capabilities. This study used a survey among ISO 9001 certified manufacturing corporations and Smart PLS Version 3.0 was used to analyze the data. The results of the study identified that process design and process control practices on the whole, had a significant relationship towards their corporate sustainable development. Also, process improvement was identified to have a significant relationship towards social development. Additionally, this study found that the effect of Process Control towards Environmental Development, Economic Development and Social Development is stronger when Knowledge Management Capabilities is higher among ISO 9001 certified manufacturing corporations in Malaysia.

Keywords: Environmental Development · Economic Development · Social Development · Sustainability

1 Introduction

With the increasing global competition faced by corporations today, without a doubt, the significance of knowledge and learning enhances the possibility of corporations to reach higher levels of Corporate Sustainable Development. This can be seen by the increasing number of corporations in Malaysia that have recognized that their Knowledge Management Capabilities are key resources for effectiveness, and a renewable resource that can

Chin, C. M. M. et al. (Eds.): SSIEC 2024, PTT, pp. 34–38, 2025.
https://doi.org/10.1007/978-981-96-2806-3_6

be utilized to accomplish optimal value and sustainable development [1]. Then again, developed countries benefit and make valuable contributions to Corporate Sustainable Development through Knowledge Management Capabilities because they have access to information strategies, up-to-date technologies and abundant resources [2]. Developing or underdeveloped countries, on the other hand, face obstacles to accessing technologies and may not have the finances to invest in further research and development.

Thus, this study seeks to recognize the moderating effect of Knowledge Management Capabilities towards Process Management Practices and Corporate Sustainable Development (refer Fig. 1 for the Research Framework). The following are the research questions for this study:

1. Is there a direct relationship between Process Design implementation towards Corporate Sustainable Development in Malaysia?
2. Is there a direct relationship between Process Improvement implementation towards Corporate Sustainable Development in Malaysia?
3. Is there a direct relationship between Process Control implementation towards Corporate Sustainable Development in Malaysia?
4. Is the relationship between Process Management Practices and Corporate Sustainable Development stronger when Knowledge Management Capabilities is higher?

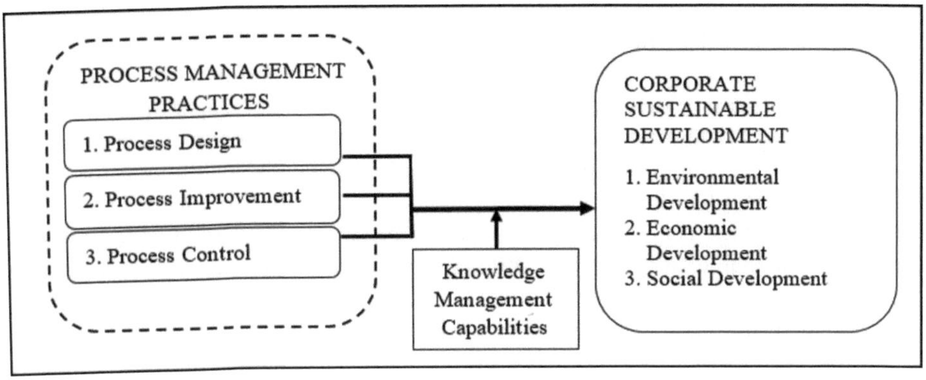

Fig. 1. Research Framework

2 Literature Review

2.1 Process Management Practices

Process Management is described as the design, control, and improvement of processes [3]. Quality gurus such as Deming [4] and Juran [5] proposed that Process Management is comprehensively advantageous to all corporations. Nowadays, the evolving type of management has identified the prominence of various processes to yield a product, especially those connected intricately to the production of a product. Therefore, this new management style places extra attention to processes [6]. The work by Aquilani et al. [7] affirms this concept by asserting the substantial effect of process standardization on overall outcomes, keeping in mind the completion time, cost and process quality. Nguyen et al. [8] approves of this notion and establishes that a corporate's processes must be improved to improve overall corporate performance. In this study, Process Management Practices is examined as the independent variable towards Corporate Sustainable Development.

2.2 Knowledge Management Capabilities

An increasing number of corporations have started transforming to highly sustainable ones [9]. To attain their sustainability goals, Knowledge Management Capabilities may be fundamental in any corporation [10]. There are numerous viewpoints as to how Knowledge Management Capabilities can be defined. One applied definition of Knowledge Management Capabilities is to view it as an efficient manner of creating, sharing and leveraging knowledge within and outside corporations [11]. Presently, the most vital point in Knowledge Management Capabilities is making sure that available knowledge in a corporation is utilized extensively to increase the productivity of operational processes in the corporation [12].

In most existing literatures on Knowledge Management Capabilities, corporations believe that Knowledge Management Capabilities will be applied effectively once created [11]. However, just the identification and distribution of important knowledge within a corporation does not guarantee a successful implementation and that the expected outputs will result when applied within the operations of a corporation [13]. Therefore, it is important to measure the impact of Knowledge Management Capabilities based on ongoing Process Management Practices in a corporation and thereafter determine the benefits that can be reaped via fitting Knowledge Management Capabilities efforts [14]. In line with inconsistent findings from literatures, this study utilizes Knowledge Management Capabilities as the moderating variable due to its inconsistent nature towards Corporate Sustainable Development. Based on past literatures noted above, the nature of the moderator Knowledge Management Capabilities is proposed to positively affect and strengthen the relationship of Process Management Practices towards Corporate Sustainable Development.

3 Methods

The survey was administered to all ISO 9001 certified manufacturing corporations listed in the Federation of Malaysian Manufacturers Directory which consists of 996 certified corporations. The unit of analysis for this research is organization, where in each organization, the respondents were either the respective human resource manager, operations manager, executive or other fitting officer in the corporation. After the 3 months specified time frame, 219 questionnaires were attained that sums to 21.9% response rate. Of all the questionnaires attained, 6 questionnaires were rejected because it had missing data higher than 50% of the survey items. Hence, 213 responses were identified to be fit for the data analysis process. Smart PLS Version 3.0 was used to analyze the data.

4 Results and Discussion

The analysis indicated that Process Design showed a positive relationship with Environmental Development, Economic Development and Social Development. The reason for this result is because corporations are adopting Process Design for the improvement of their sustainability and corporate performance right from their designing process and planning stage. Similarly, Movahedi et al. [15] stated that, corporations have started using eco-friendly raw materials and emphasizing on pollution reduction right from the stage of designing their processes to reducing overhead costs and to display their commitment towards safeguarding the environment and society at large.

Additionally, this study identified a positive relationship between Process Improvement and Social Development. This finding could be due to the continuous demanding consumers that require corporations to increase the improvement initiatives in their operational processes to address sustainable issues. According to Lopes et al. [16], corporations are showing keen interest in Process Improvement activities to achieve greater efficiencies and improved production levels.

Next, Process Control was identified to have a significant relationship towards all the dimensions of Corporate Sustainable Development. The possible reason for this situation could be because Process Control is sometimes used in corporations to monitor the environmental, economic and social reach of the operational processes that optimized sustainability and efficiency. This finding is supported by Thomas et al. [17], where the implementation of Process Control enables corporations to indulge in green output or sustainable productions that leads to de-creased energy usage, pollution and waste management and resource depletion for their Corporate Sustainable Development.

This study found that the effect of Process Control towards Environmental Development, Economic Development and Social Development is stronger when Knowledge Management Capabilities is higher among ISO 9001 certified manufacturing corporations in Malaysia. The finding of this study is supported by Recker and Mendling [18], whereby many manufacturing corporations place process controls, and boost their operations by transforming into a knowledge intensive learning corporation.

5 Conclusion

The results of the study identified that Process Design and Process Control practices on the whole, had a significant relationship towards Corporate Sustainable Development. Also, Process Improvement was identified to have a significant relationship towards Social Development. Additionally, this study found that the effect of Process Control towards Environmental Development, Economic Development and Social Development is stronger when Knowledge Management Capabilities is higher among ISO 9001 certified manufacturing corporations in Malaysia. In conclusion, Process Management Practices are not an additional burden on corporations but provide significant benefits. The results of this study should motivate managers to adopt Process Management Practices to achieve higher levels of Corporate Sustainable Development. In addition, taking advantage of Knowledge Management Capabilities can improve business processes, which ultimately leads to greater efficiency and effectiveness.

Acknowledgments. The authors would like to express their appreciation to Universiti Sains Malaysia Research University (RU) Grant (1001/PMGT/8016100) for the financial support to carry out this research.

References

1. Salwa, H.A., Sakundarini, N., Raja, A.R.G., Ramayah, T.: The impact of sustainable manufacturing practices on sustainability performance: empirical evidence from Malaysia. Int. J. Oper. Prod. Manag. **37**(2), 182–204 (2017)
2. Al-Roubaie, A., Alvi, S.: Knowledge transfer for sustainable development: east-west collaboration? World J. Sci. Technol. Sustain. Dev. **11**(4), 242–255 (2014)
3. Jones, J.L.S., Linderman, K.: Process management, innovation and efficiency performance: the moderating effect of competitive intensity. Bus. Process. Manag. J. **20**(2), 335–358 (2014)
4. Deming, W.E.: Out of the Crisis. MIT Center for Advanced Engineering Study, Cambridge, MA (1986)
5. Juran, J.M.: The quality trilogy: a universal approach to managing for quality. Qual. Prog. **19**, 19–24 (1986)
6. Dave, B.: Business process management –a construction case study. Constr. Innov. **17**(1), 50–67 (2017)
7. Aquilani, B., Silvestri, C., Ruggieri, A.: Sustainability, TQM and value co-creation processes: the role of critical success factors. Sustainability **8**(10), 995 (2016)
8. Nguyen, M., Phan, A., Matsui, Y.: Contribution of quality management practices to sustainability performance of Vietnamese corporations. Sustainability **10**(2), 375–390 (2018)
9. Gao, J., Nee, A.Y.C.: An overview of manufacturing knowledge sharing in the product development process. Proc. Inst. Mech. Eng. Part B J. Eng. Manuf. **32**(13), 2253–2263 (2018)
10. Costa, V., Monteiro, S.: Key knowledge management processes for innovation: a systematic literature review. VINE J. Inf. Knowl. Manage. Syst. **46**(3), 386–410 (2016)
11. Evangelista, P., Durst, S.: Knowledge management in environmental sustainability practices of third-party logistics service providers. Vine **45**(4), 509–529 (2015)
12. Aboelmaged, M.G.: Linking operations performance to knowledge management capability: the mediating role of innovation performance. Prod. Plan. Control **25**(1), 44–58 (2014)
13. Breite, R., Koskinen, K.U.: Supply chain as an autopoietic learning system. Supply Chain Manage. Int. J. **19**(1), 10–16 (2014)
14. Wang, Z.N., Wang, N.X.: Knowledge sharing, innovation and firm performance. Expert Syst. Appl. **39**(10), 8899–8908 (2012)
15. Movahedi, B., Miri-Lavassani, K., Kumar, U.: Operational excellence through business process orientation. TQM J. **28**(3), 467–495 (2016)
16. Lopes, C.M., Scavarda, A., Hofmeister, L.F., Thomé, A.M.T., Vaccaro, G.L.R.: An analysis of the interplay between organizational sustainability, knowledge management, and open innovation. J. Clean. Prod. **142**, 476–488 (2017)
17. Thomas, A., Byard, P., Francis, M., Fisher, R., White, G.R.T.: Profiling the resiliency and sustainability of UK manufacturing corporations. J. Manuf. Technol. Manag. **27**(1), 82–99 (2016)
18. Recker, J., Mendling, J.: The state of the art of business process management research as published in the BPM conference: recommendations for progressing the field. Bus. Inf. Syst. Eng. J. **58**(1), 55–72 (2016)

Additive Manufacturing in the Aerospace Industry: Malaysia versus Global Market Outlook

Syed Ahmad Faiz Syed Kamarulzaman[1], Novita Sakundarini[2](✉), Chin Seong Lim[2], and Wee King Law[2]

[1] Aerospace Malaysia Innovation Center, Kajang, Selangor, Malaysia
[2] Faculty of Science and Engineering, Department of Mechanical, Material and Manufacturing Engineering, University of Nottingham Malaysia, Semenyih, Selangor, Malaysia
novita.sakundarini@nottingham.edu.my

Abstract. Additive manufacturing (AM) refers to manufacturing technologies which feature the joining of materials in a layer wise fashion to make 3-dimensional (3D) components using computer-aided design (CAD) models as input. Additive Manufacturing is a growing technology, and its growth has been gaining industry's attention in the last decade. There has been a steep incline on both the AM innovation curve which forms the supply and the amount of AM industrial applications which represents the demand. These increases have been observed overseas, whereas the general perception of AM's activity in Malaysia is that it is still in its infancy. Therefore, this research aims to understand the current Malaysian outlook in comparison with global market outlook, and specifically in the aerospace industry.

Keywords: Additive Manufacturing · 3D Printing · Global Market Outlook · Aerospace Industry

1 Introduction

Market research is a crucial aspect of this study for the industry players to fully understand the current state of the additive manufacturing ecosystem and its potential future in Malaysia. This can be done by breaking it into different subcomponents to gain a better understanding of the overall AM ecosystem. The component of market research used in this study consists of elements as shown in Fig. 1.

The research aims to answer the following research questions:

RQ1 What is the state of AM Malaysia market outlook

RQ2 What is the AM market outlook for Malaysia aerospace industry in comparison with AM global market outlook?

The present study aims to provide insight into current AM practices implemented by Malaysian industries and give recommendations on the possible improvements to a greater extent of these practices. The present study can facilitate manufacturing companies toward adopting AM to remain competitive in the future as there has been very

Chin, C. M. M. et al. (Eds.): SSIEC 2024, PTT, pp. 39–44, 2025.
https://doi.org/10.1007/978-981-96-2806-3_7

Fig. 1. Components of market research used in this study.

limited research undertaken on AM implementation in Malaysia [1]. Therefore, the purpose of the present study is mainly to obtain new insights about this subject and is classified as 'exploratory' in nature.

2 Methodology

2.1 Research Phase

This study used a mixed method of quantitative and qualitative study. The 'industrial wish list, readiness and challenges', and the 'local activities' components will be broken down into 3 levels of assessment:

1. **Level 0 (L0)** is the engagement with AM industrial players (hardware, software, material manufacturers and suppliers, both local and foreign entities which has a business footing in Malaysia). L0 engagements will provide a basic understanding of the supply available in Malaysia as well as the perceived demand by the industrial end user. Further to that, L0 will provide an understanding of the local industrial use and adoption of AM by the local market. Finally, L0 can provide insight into the strength and sustainability of the local AM ecosystem. Seventeen participants are involved in this level consisting of local distributors of overseas AM solutions, and their companies were Malaysian fully owned, small to medium enterprises (SMEs).
2. **Level 1 (L1)** is a multi-industry engagement, which is made up of four industries which are aerospace, automotive, oil & gas as well the electronics industry's support services sector. The latter three industries were chosen due to the size of their presence in Malaysia as well as a prior understanding of AM use cases within these industries. L1 will provide an understanding of the awareness, interests, and requirements for AM and identification of any ongoing AM-related activities within these industries. L1 will also be able to give an indicator of the need for AM within the local industries. The justification for this level is to develop a comparative understanding of the aforementioned items between aerospace and other industries. Total participants engage in this survey are 56 participants.
3. **Level 2 (L2)** is the multi-sector, single-industry engagement focused on the aerospace industry. Malaysia's aerospace industry can be broken down into 6 subsectors namely:
   - Aerospace manufacturing
   - Maintenance, repair & overhaul (MRO)

- Research & technology
- Education
- Engineering services
- Systems Integration

L2 will be focused on understanding the needs, readiness and challenges of the aerospace industry in adopting AM within their organization. At this level, any adoption or use of AM is of great interest and requires a more in-depth understanding of the journey that the organization has taken to decide on their use or adoption of AM. L2 will provide a good understanding of the industrial problems within the aerospace industry that can be solved using AM which will then allow for the identification of AM's niche area. The total participants of this level are 33 respondents. Figure 1 shows the complete research methodology used in this study.

2.2 Surveys Development

The present study used a mixed method of quantitative and qualitative. The quantitative approach is used to understand the awareness of AM, its use, and its adoption trends, and to assess the local AM ecosystem. This approach is reflected in the questionnaires that use closed-ended questions measured on a Likert scale (Table 1).

Table 1. Summary of the relation between objectives and methods.

Objectives	Survey method used
1. Assess AM awareness amongst Malaysia's industry players	Questionnaire
2. Assess AM use and adoption trends	Questionnaire
3. Assess the local AM ecosystem	Questionnaire
4. Identify growth factors, requirements, and challenges for AM implementation in Malaysia	Questionnaire, interview, focus group discussion
5. Identify AM's niche within the local context	Questionnaire, interview, focus group discussion

The qualitative approach is used to identify growth factors, requirements and challenges faced by industries on the implementation of AM. This is reflected in the questionnaires that used open-ended questions, interview sessions and group discussions focused on.

2.3 Pilot Test of Instruments

To ensure the validity of the research instrument used in the present study, a pilot test is conducted as follows:

1. A pilot survey using the designed instrument is conducted on 5 professionals and four academics to get their feedback on how they can understand the questions.

2. Following the pilot survey, the questionnaire is refined according to the participants' comments (Fig. 2).

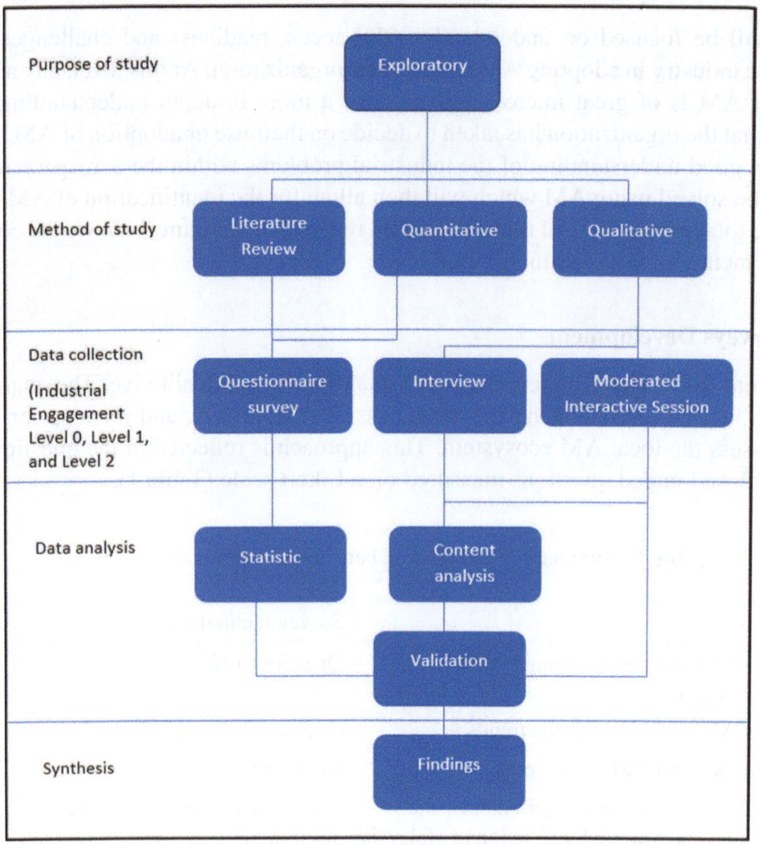

Fig. 2. Methodology of study.

3 Results and Discussion

AM usage in the aerospace sector in Malaysia is mainly for R&D purposes, prototyping, repair, jig, fixtures, and tooling which is agreeable with the overall results from all industries. However, in the aerospace industry locally, there is no application for production or end-use products. The AM technology used in aerospace industries is Material Extrusion and Vat Polymerisation which is parallel with the overall results. In terms of AM usage experiences, only seven out of fifteen aerospace companies indicated that they have experience in using AM technology for either prototyping, repair or jig, fixtures, and tooling.

The following section will discuss the comparison on local versus the global outlook of AM which focuses on the use, applications and its barriers.

- Usage of Various AM Technologies in Malaysia and Overseas
  It can be inferred that Malaysia's AM ecosystem is currently dominated by polymer AM technologies such as Material Extrusion and Vat Photopolymerisation. This is evidenced by the high usage rankings as reported by Malaysia's AM vendors and industry players. Furthermore, both parties agreed that there is significant demand for 3D printing services as well as AM implementation from Malaysia's industry players.

  The usage trend of various AM technologies was found to agree with those reported in research papers as well as industrial surveys. A survey by Hubs reported that the use of AM technologies has increased tremendously in recent years. This resulted in significant revenue growth (about 15.1 billion USD) and was predicted to increase about three-fold in 2026 [2]. With this growth, AM will be amongst the forefront fabrication method in the future.
- Applications of AM in Malaysia and Overseas
  Malaysia's AM vendors and industry players generally agreed that the three main applications of AM include the fabrication of prototypes, production parts, as well as tooling. Furthermore, Malaysia's industry players responded by saying that current AM applications were more focused on research and development as compared to prototyping and fabrication of production parts. A similar trend was identified when comparing the applications of AM in Malaysia with those from overseas. The three main applications of AM were identified as research and development, followed by prototyping, and subsequently fabrication of production parts [3]. It could be inferred that this trend will continue into the near future until AM technologies reach a certain level of maturity in terms of research and development. Following that, the trend may gradually shift towards the fabrication of production parts due to it being mature enough for implementation in the industry.
- Barriers to AM adoption in Malaysia and overseas
  A similar trend was identified when comparing the barriers to AM adoption in Malaysia with those from overseas. Qualification and certification were ranked as the main barriers to AM adoption, followed by high upfront costs for 3D printers, and subsequently lack of in-house expertise in 3D printing [4]. Thus, it can be inferred that the successful implementation of AM technologies in a particular industry is directly related to the establishment of proper qualification and certification standards for AM technologies and the parts fabricated by them. With these standards in place, industry players now possess clear guidelines for fabricating 3D-printed parts that comply with industrial standards, in addition to fulfilling application requirements.

4 Conclusion

The global adoption of AM across major industries is continuously growing and the growth of end parts is increasing. Locally, the interest in using AM is high, but awareness of AM's full capability and technical know-how is low. The technological maturity and level of industrialization for some AM technologies have also reached a high level in general. The ecosystem in Malaysia can be said to be in its infancy stage. A lot more growth is required for the ecosystem to be sustainable. The current ecosystem needs to grow but also for growth to happen, business content must exist.

Malaysian Aerospace players (particularly aero manufacturing, MRO & systems integration) should explore and use AM on their shopfloor for manufacturing in-house manufacturing aids followed by complex fixtures and tooling as these parts do not require rigorous certification. This will show OEMs and Tier 1 companies the willingness of local companies to use the latest technologies and may help them capture potential future business. At this stage, companies can leverage the use of machines by 3D printing services bureaus with the collaboration of universities to validate the fabrication process and conduct material testing. The speed at which this happens will be critical due to the speed of AM uptake in the global aerospace field.

Acknowledgements. We would like to thank Aerospace Malaysia Innovation Center and University of Nottingham Malaysia for their support on the Additive Manufacturing for Aerospace Tools and Components (AMATOC) project, especially during the market research study.

References

1. Campbell, T., Williams, C., Ivanova, O., Garrett, B.: Could 3D Printing Change the World? Technologies, Potential, and Implications of Additive Manufacturing, (2011)
2. Guo, N., Leu, M.C.: Additive manufacturing: technology, applications and research needs. Front. Mech. Eng. **8**, 215–243 (2013). https://doi.org/10.1007/s11465-013-0248-8
3. Attaran, M.: The rise of 3-D printing: the advantages of additive manufacturing over traditional manufacturing. Bus. Horiz. **60**, 677–688 (2017). https://doi.org/10.1016/j.bushor.2017.05.011
4. Khorram Niaki, M., Nonino, F.: The Management of Additive Manufacturing. Springer, Cham (2018). https://doi.org/10.1007/978-3-319-56309-1

Noise and Errors Identification in Retail Checkout Systems: A Survey-Based Study

Shazielya Shamsul[1] (✉), Salwa Hanim Abdul-Rashid[1,2], and Mohammed A. H. Ali[1,2]

[1] Department of Mechanical Engineering, Universiti Malaya, Kuala Lumpur, Malaysia
shazielyaworks@gmail.com

[2] Centre for Sustainable and Smart Manufacturing, Universiti Malaya, Kuala Lumpur, Malaysia

Abstract. In the realm of modern retail, supermarkets play a vital role as a one-stop destination for purchasing essential goods, offering a wide range of products under one roof. However, this convenience is often compromised by persistent issues arising from operational errors, resulting in customer dissatisfaction and delays. The existing technology employed in supermarket checkout counters faces several challenges, including issues related to both barcode and Radio Frequency Identification (RFID) technology. This paper aims to identify the noise, and errors associated with both barcode and RFID technologies in retail checkout systems through a survey-based study, with implications for enhancing the overall shopping experience and optimizing retail operations. An online survey-based study was conducted to gather customers' views on both barcode and RFID systems. The findings indicate that customers prefer self-checkout using RFID technology over barcode technology due to its improvement in terms of time, speed, and efficiency.

Keywords: Noise and Errors · Barcode · Radio Frequency Identification (RFID) Technology

1 Introduction

1.1 Background

The shift from face-to-face interactions to digital communication has significantly impacted the way people live their daily lives, including the way they shop. Despite the rise of e-commerce, many customers still prefer to shop in physical stores, valuing the tactile experience of physically examining and touching products. According to some studies, 51% of shoppers prefer to purchase online, while 49% prefer to visit the store [1]. However, the checkout process can be a laborious and time-consuming experience, especially during peak hours or when there are limited checkout lanes available. This inefficiency can lead to frustration and dissatisfaction for both customers and staff, contributing to lost sales as impatient customers may abandon their purchases and leave the store altogether.

Chin, C. M. M. et al. (Eds.): SSIEC 2024, PTT, pp. 45–50, 2025.
https://doi.org/10.1007/978-981-96-2806-3_8

The current RFID system has the potential to streamline the checkout process by enabling customers to place all their items into a single smart RFID basket, offering a convenient and user-friendly experience for people of all ages. However, the basket may sometimes fail to detect specific RFID tags, especially those related to pricing information, or it may inaccurately register an item multiple times. When these errors occur, customers often require assistance from staff, which can disrupt their shopping experience. This can lead to long queues and delays at the checkout counter, causing frustration and disappointment for many customers. Moreover, it has been reported in literature that common problems in current RFID systems include RFID tag readability issues, where tags may not be accurately detected or read due to environmental factors, hardware limitations, or other technical constraints [2, 3]. Furthermore, tags may be lost or damaged, which can exacerbate the issues with RFID tag readability [4]. The system may be affected by multiple frequencies of tags, which can cause interference and disrupt the reading process, further compromising the reliability of the RFID checkout system [5, 6].

2 Methodology

The survey aims to identify the noise and errors that may affect the checkout experience. The online survey consisted of two types of questionnaires: (1) customers who have used barcode systems with cashiers or self-service checkout counters, and (2) customers who have used RFID systems.

2.1 Online Survey Questionnaire

The first questionnaire was created to collect feedback from customers who have used barcode systems at retail stores, either with cashiers or through self-service checkout counters. It focused on various aspects of the checkout experience, such as the time required for the payment process, common issues faced by cashiers (including challenges, speed, and errors related to the barcode system), and customer satisfaction levels with the current system. The goal was to provide a comprehensive understanding of the pros and cons of the barcode system from the customer's perspective.

The second questionnaire aimed to gather insights from customers who have used the RFID system at retail stores. It sought to identify factors affecting customer preferences and experiences with the RFID system, including the time needed to complete the payment process, satisfaction levels, problems or errors encountered, and overall perception of the RFID system. This self-service system differs from barcode systems as it does not require cashiers to scan items, offering a unique experience. The study aimed to provide a detailed understanding of customer preferences and experiences with different checkout technologies. The findings can help retailers enhance their checkout processes and overall customer experience. The location of the data collection for the online survey of the current checkout systems are illustrated in Fig.1.

Fig. 1. Location of the data collection: (a) Barcode system at the supermarket, and (b) RFID system at a sport retail store.

3 Result and Discussion

The study analysed online survey questionnaires to gain insights into the behaviours, interactions, and preferences of the target audience. The results provided valuable information on what participants want and expect from barcode and RFID checkout systems in retail stores.

3.1 Barcode Systems

According to the data, price is the most significant factor in customers' purchasing decisions, with 86.1% (198 out of 230) considering it their top priority (Fig. 2(a)). Customers also value a quick checkout, with 67% (154 out of 230) expecting the process to take 3 to 5 min (Fig. 2(b)). However, 55.9% and 28.4% of customers experience long queues and wait times, respectively, during checkout (Fig. 2(c)). Additionally, issues such as undetectable barcodes, missing or inaccessible barcodes, and improperly placed barcode labels are common (Fig. 2 (d), (e), and (f)). Notably, 58.1% (133 out of 229) cite missing or inaccessible barcodes as the primary cause of delayed scanning times, underscoring the need to improve checkout efficiency.

Next, 80.4% of customers (185 out of 230) see the checkout process as a major area for improvement, while 77.8% (179 out of 230) consider reducing waiting times a key concern (Fig. 3(a)). Fig. 3(b) shows that most customers rate the current barcode system's performance as a moderate 3 out of 5, reflecting a moderate level of satisfaction. Additionally, 51.5% (117 out of 227) of customers rate the system's effectiveness in facilitating the payment process as 3 out of 5, as shown in Fig. 3(c).

3.2 RFID Systems

Figure 4 (a) shows that 80.2% of customers (162 out of 202) had previously visited a sports retail store in Malaysia, while 19.8% (40 out of 202) had yet to visit the store. According to the Fig. 4(b), a total of 56% (114 out of 203) of the customers has previously used the RFID system, giving them a firsthand understanding of its capabilities and limitations. Notably, 36.9% of customers who had used the system (73 out of 162) gave it the highest rating of 5 out of 5, indicating an extremely high level of satisfaction with their experience using the system, as demonstrates in Fig. 4(c).

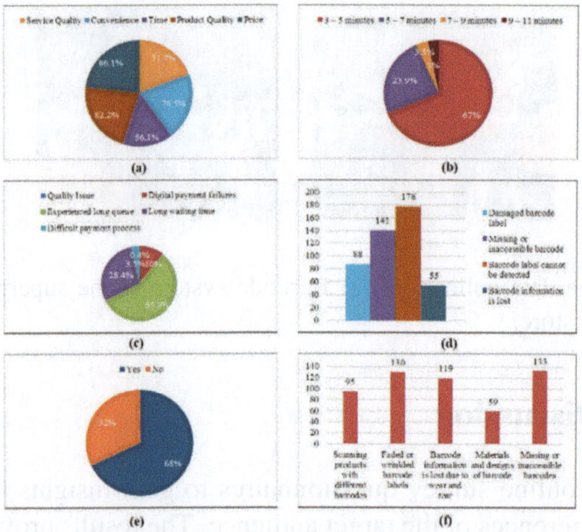

Fig. 2. Survey Results of Barcode System: (a) Values considered when shopping; (b) Average time required for payment process; (c) Primary issues encountered at checkout counter; (d) Most common issues: Cashier challenges; (e) Barcode placement, and (f) Identification of the problem areas.

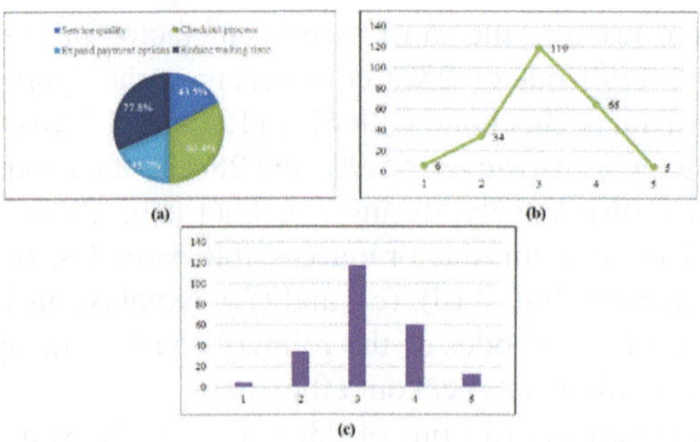

Fig. 3. Survey Results on Barcode System: (a) Areas for improvement; (b) Levels of satisfaction with the current barcode system, and (c) Level of satisfaction with the current barcode system in facilitating the payment process.

Moreover, the majority of the customers who used the RFID system experienced difficulties and errors during the payment process, as shown in Fig. 4(d). The most common problems encountered were related to RFID tag detection failure, which accounted for 62% of all reported issues (11 out of 179). Most of the customers 32.8% (63 out of 192) rate the assistance that they have received from sport retail store employees as a 3 out of 5, indicating an average level of satisfaction with the support provided, as illustrated in Fig. 4(e). The analysis of data from Fig. 4(f) shows that the majority of customers who used the RFID system at a sport retail store found the experience of placing their shopping items into the boxed area to be satisfactory, with an average rating of 3 out of 5. This corresponds to 40.6% of the customers (78 out of 192), indicating that most of them were generally satisfied with their overall experience at the self-checkout system, despite some issues they may have encountered during the payment process.

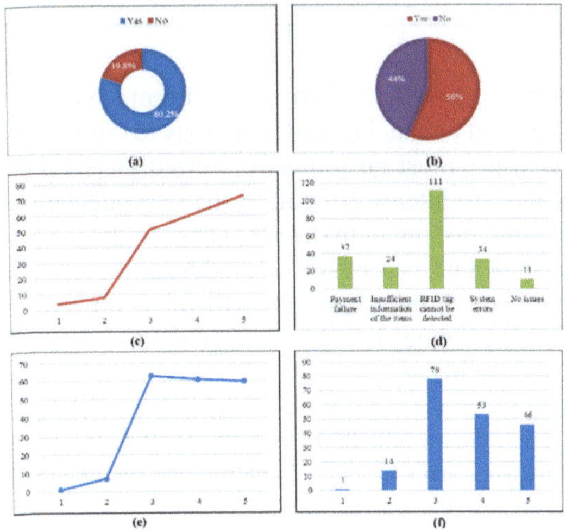

Fig. 4. Survey Results on RFID System: (a) Frequency of visits; (b) Analysis of user interaction with the RFID system; (c) Levels of satisfaction; (d) Problems or errors encountered; (e) Customer satisfaction ratings for employee assistance, and (f) Ratings on the difficulty of placing shopping items into boxed area.

According to the data presented in Fig. 5(a), the majority of customers (38.3%) anticipate spending around 2-3 min on the payment self-checkout process using the RFID system. Additionally, as shown in Fig. 5(b), a significant proportion of customers (35%) or (69 out of 195) gave a perfect score of 5 for the RFID system's ability to facilitate the payment process, indicating a high level of satisfaction with its functionality. Furthermore, 42.1% of customers would recommend the system to a friend or colleague, suggesting a strong level of customer satisfaction and confidence as demonstrates in Fig. 5(c). Overall, customer feedback indicates that most users have a positive experience, with 35.8% giving a rating of 4 out of 5, as illustrated in Fig. 5(d).

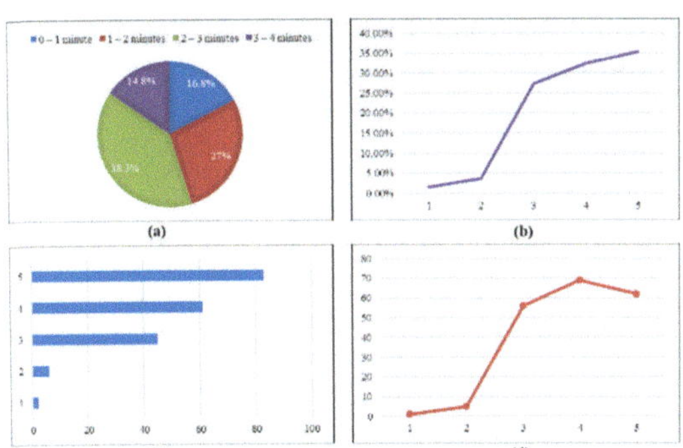

Fig. 5. Survey Results on RFID System: (a) Average time required; (b) Customer perception of the RFID system; (c) Likelihood of recommending RFID system to friends or colleagues, and (d) Customer ratings of the RFID system.

4 Conclusion

The study aims to identify the noise and errors in current checkout systems through an online survey, with the goal of enhancing the technology to improve customer experience. The results reveal that while most customers preferred the RFID-based self-checkout system, some individuals encountered difficulties such as undetected RFID tags, payment failures, system errors, and missing item details. These findings underscore the need for an upgraded RFID system that can effectively address these issues and enhance the overall customer experience. Future research will focus on developing an intelligent RFID system that can streamline the checkout process and provide a seamless and efficient experience for customers.

Acknowledgements. The authors would like to thank the Centre for Sustainable and Smart Manufacturing, Universiti Malaya, Kuala Lumpur, Malaysia, for their funding and support of this research.

References

1. Setiawan, R., et al.: References for shopping online versus in stores: what do customers prefer and how do offline retailers cope with it? Prod. Manag. **1**(1), 874–898 (2020)
2. Nisha, F.: Implementation of RFID technology at defence science library, DESIDOC: a case study. DESIDOC J. Libr. Inf. Technol. **38**(1), 27–33 (2018)
3. Cheng, S.T., Shih, K.X., Cheng, H.K., Kung, C., Wang, C.Y.: Design of books inventory with RFID antenna in library management system. In: MATEC Web of Conferences, vol. 123, p. 00028 (2017)
4. Paaske, B.A., Moser, T., Seckman, C.: The benefits and barriers to RFID technology in healthcare. Online J. Nurs. Inform. OJNI; Chicago **21**(2) (2017)
5. Aboelmaged, M., Hashem, G.: RFID application in patient and medical asset operations management: a technology, organizational and environmental (TOE) perspective into key enablers and impediments. Int. J. Med. Inform. **118**, 58–64 (2019)
6. Shah, S.N., Abuzneid, A.: IoT based smart attendance system (SAS) using RFID. In: 2019 IEEE Long Island Systems, Applications and Technology Conference (LISAT), Farming-dale, NY, USA, pp. 1–6 (2019)

Investigation of Autonomous Mobile Robot Path Planning with Edge Cloud Based on Ant Colony Optimization

Siti Nur Lyana Karmila Nor Azmi[1(✉)], Majid Rafique[1], Nur Ilyana Anwar Apandi[1], and Nor Aida Zuraimi Md Noar[2]

[1] Faculty of Electrical Technology and Engineering, Universiti Teknikal Malaysia Melaka, Durian Tunggal, Malaysia
M112310028@student.utem.edu.my

[2] Faculty of Science and Mathematics, Universiti Malaysia Pahang Al-Sultan Abdullah, Pekan, Malaysia

Abstract. Autonomous Mobile Robots (AMRs) have become increasingly significance across various industries, with efficient path planning being a crucial component for their operation, particularly in dynamic and unpredictable environments. Although extensive research has been conducted on cloud-based robotics path planning, significant challenges remain in addressing the latency and adaptability required for real-time decision making. This paper presents a novel approach that integrates edge cloud-computing with an improved Ant Colony Optimization (ACO) algorithm to overcome these challenges. By decentralizing computational task to the edge, the proposed method not only reduces latency but also improved the speed and accuracy of route planning. The integration of a global path storage system further accelerates the process by enabling to reuse the optimized paths in the future planning scenarios, thus improving efficiency. Moreover, the proposed ACO algorithm introduces a reduced pheromone deposit strategy, which accelerates convergence and reduces the number of iterations needed to find the optimal path. Simulation results demonstrate that the proposed approach significantly outperforms conventional algorithms in terms of both adaptability and convergence speed, leading to shorter and more efficient paths in complex, dynamic environments. These findings suggest that this collaborative system of edge-cloud and ACO approach could be instrumental in advancing the deployment of AMRs in industrial applications where real-time path planning and adaptability are critical.

Keywords: Autonomous Mobile Robot (AMR) · Ant Colony Optimization (ACO) · Path Planning · Edge Cloud Computing · Dynamic Environment

1 Introduction

Autonomous Mobile Robot (AMR) are increasingly important in various industries, from manufacturing and logistics to healthcare and service sectors. These robots are designed to navigate autonomously within an environment, making decisions in real-time to avoid obstacles and reach their destinations efficiently [1, 2]. The effectiveness

Chin, C. M. M. et al. (Eds.): SSIEC 2024, PTT, pp. 51–56, 2025.
https://doi.org/10.1007/978-981-96-2806-3_9

of AMRs in these dynamic and unpredictable environments heavily depends on robust path planning algorithms. However, most existing studies have primarily focused on static environments, leaving a significant gap in addressing the challenges of dynamic settings where both the environment and obstacles can change unpredictably.

Among the various algorithms developed for path planning, Ant Colony Optimization (ACO) has emerged as a promising solution due to its bio-inspired approach to problem-solving. ACO is a population-based metaheuristic algorithm inspired by the natural foraging behavior of ants. In nature, ants find the shortest path between their nest and a food source by laying down pheromone trails, which, over time, guide more ants to the optimal path as the pheromone concentration increases. In computational terms, ACO mimics this behavior to solve optimization problems, particularly those involving pathfinding [3]. For AMR path planning, ACO deploys virtual ants in a simulated environment to explore different routes from a start point to a destination, dynamically updating a pheromone matrix based on the paths taken [4]. The pheromone evaporation rate ensures that paths become less attractive over time unless they are reinforced, encouraging the exploration of potentially better routes [5].

Despite its advantage, classic ACO algorithms face significant challenges, particularly in dynamic environments where obstacles may appear or disappear unpredictably This limitation is compounded by the slow convergence speed of ACO, which can delay the discovery of optimal paths, especially in scenarios requiring real-time decision making [4]. Furthermore, the computational expense of continually updating the pheromone matrix and analyzing numerous pathways at once can be significant, especially in broad and complex settings.

Thus far, most research has initially focused on path planning in static environments, where obstacles remain static and the robot is the only mobile entity, as opposed to real-world scenarios where obstacles may also be mobile. One notable example of this is the study referenced in [6], which focuses on path planning implementing ACO and edge cloud computing but within static environment. The dynamic nature of real-world scenarios requires an approach that not only adapts to changing environments but also converges quickly to optimal solutions.

Edge cloud computing emerges as a potential solution to these issues by bringing computational resources closer to the source of data generation, thus reducing the reliance on centralized cloud servers [7]. In traditional cloud computing models, data generated by devices like AMRs is sent to a centralized server for processing, which can introduce latency due to the time required to transmit data to and from the cloud. This latency is particularly problematic in dynamic environments where real-time decision-making is critical, hindering AMR performance. By decentralizing computational tasks, edge cloud computing allows path planning algorithms like ACO to be executed on edge devices, significantly reducing latency and enabling quicker responses to environmental changes. This paper will explore how the integration of edge computing with an optimized ACO algorithm can overcome the challenges, improving both the adaptability and convergence speed of AMR path planning in dynamic environments.

2 Methodologies

In this section, we present the approach that integrates edge cloud computing with an optimized ACO algorithm to improve both the adaptability and convergence speed of AMRs in dynamic environments. As illustrated in Fig. 1, the proposed system architecture is designed to utilize the decentralized computing power of edge cloud systems to improve the performance of AMRs in dynamic environments. The architecture is divided into two main components which the **Edge** and the **Cloud.**

Based on Fig. 1 below, the AMR gathers environmental data at the edge and uses the ACO algorithm to explore potential paths. This initial computation is then sent to the cloud, where the data is further processed, incorporating prior knowledge and detecting any potential issues such as loop formations. The cloud component is also responsible for storing optimized paths, which can be retrieved and reused in future scenarios, significantly reducing the computational load and improving convergence speed.

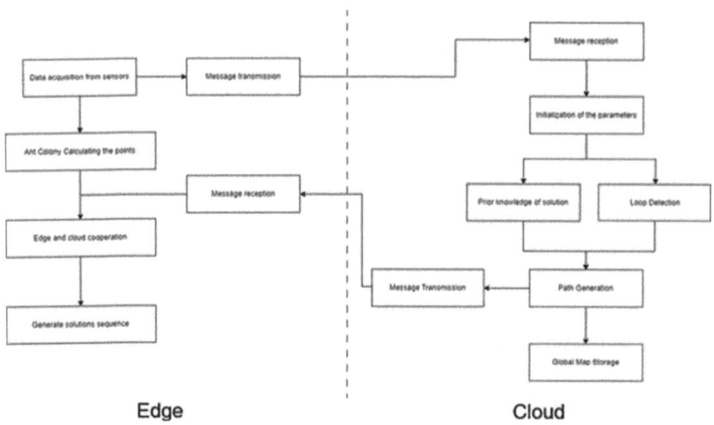

Fig. 1. The edge-cloud computing integrative ACO path planning system architecture

As for the environmental design, this system operates within a 30 × 30 grid map that as shown in Fig. 2(a), which serves as the environment for AMR navigation. This grid-based model represents the workspace of the robot, where each cell in the grid can either be occupied or unoccupied. The environment is designed to be dynamic, with obstacles that can appear or disappear, simulating real-world conditions where the AMR must adapt to changing surroundings.

This grid is critical for testing the adaptability and efficiency of the ACO algorithm. The main task of the AMR is to navigate from a starting point to a destination within this grid, dynamically adjusting its path as obstacles change based on the mathematical formulation outlined in [8].

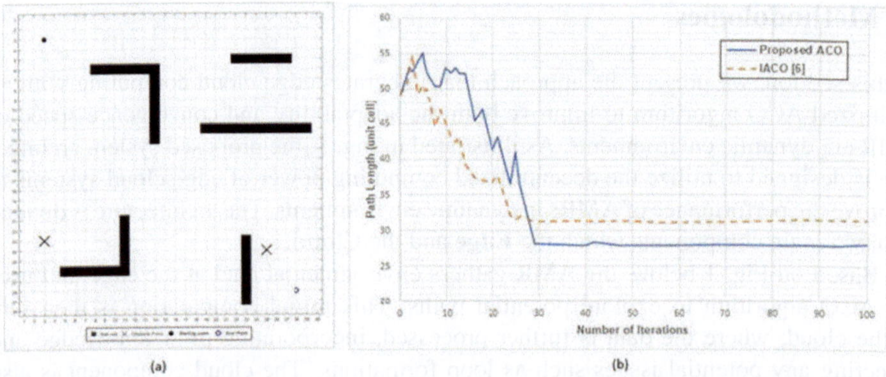

Fig. 2. (a) The 30 × 30 grid map model (b) The iterative process between IACO algorithm and Proposed ACO

In this study, the ACO algorithm follows the classic principles outlined in [6]. According to the paper, the pheromone update rule, which determines how pheromone levels are adjusted based on the path taken, which is important in guiding the search process. The probability of moving from one node to another is influenced by both the pheromone levels and heuristic information, which help in determining the most optimal paths [6]. As this approach will be in a 30 × 30 dynamic grid map, in this setup, the ACO algorithm is tasked with navigating from a starting point to a destination while continuously adapting to changes in the environment by reducing the amount of pheromone deposited on paths. This strategy prevents the algorithm from prematurely converging on suboptimal paths, which is a common issue that slows down the convergence process. By encouraging broader exploration of alternative routes, the algorithm can identify optimal paths more quickly. The path retrieval mechanism is highlighted as a critical improvement by storing the previously optimized paths and retrieving them when similar scenarios are encountered, the algorithm avoids the need to recalibrate paths from scratch. This reuse of paths significantly speeds up the convergence process.

3 Results and Discussion

This section presents the results of the simulation experiments conducted to evaluate the performance of the proposed ACO algorithm in the dynamic 30 × 30 grid environment as shown in Fig. 2(a) The performance of the ACO algorithm was evaluated based on the number of iterations, comparison analysis on the path length that was generated, and the time taken to running the process of the system. This study also was compared to IACO in [6].

As shown in Fig. 2(b), the proposed ACO algorithm required 29 iterations to converge to the optimal path, whereas the IACO required 33 iterations. This reduction in iterations demonstrates a more efficient convergence process in the proposed method. Furthermore, Table 1 presents a comparison of the algorithms in terms of shortest path length and running time. The proposed ACO achieved a shorter path length of 26.782 units compared to IACO's 31.351 units. Additionally, the running time for the proposed ACO was 9.632 s, which is faster than IACO's 11.258 s. This indicates that the proposed approach is not only more efficient in terms of pathfinding but also faster in execution.

The path retrieval mechanism significantly contributed to the improved convergence speed observed by reusing previously optimized paths. This reuse of paths reduced the need for recalculating routes from scratch, leading to faster convergence. The efficiency gain from this approach is highlighted by the fact that the proposed ACO algorithm converged approximately 12.12% faster than the IACO algorithm, as indicated by the reduction in the number of iterations and the overall computational effort. Also, along with the reduced pheromone deposit strategy, allowed the algorithm to maintain a balance between exploration and exploitation. This balance prevented premature convergence on suboptimal paths and enabled the algorithm to find the most efficient routes even in complex, dynamic settings.

Table 1. Comparison of performance of the algorithms

Algorithm	Environment	Shortest path (cell unit)	Iterations	Running Time (s)
IACO [6]	Static	31.351	33	11.258
Proposed ACO	Dynamic	26.782	29	9.632

4 Conclusion

This study proposes employing an ACO algorithm improved with edge cloud computing to address the limitations of conventional ACO for AMR path planning. The results from the simulations provide strong evidence that the proposed ACO algorithm significantly outperforms traditional methods and the IACO in [6], even in dynamic settings. This development offers improved navigation efficiency in dynamic indoor settings, promotes wider AMR implementation in industrial applications, and advances Industry 4.0 goals. For future research, we will focus on optimizing the utilization of path retrieval to further improve computational efficiency. This will involve refining the retrieval process to ensure that only the most relevant and efficient paths are reused, reducing unnecessary computations. Additionally, the impact of path retrieval optimization on overall system performance will be evaluated in more complex environments.

Acknowledgement. The study is funded by Ministry of Higher Education (MOHE) of Malaysia through the Fundamental Research Grant Scheme, No.:FGRS/1/2022/TK07/UTEM/02/33 from Universiti Teknikal Malaysia Melaka (UTeM).

References

1. Lv, T., Zhang, J., Zhang, J., Chen, Y.: A path planning algorithm for mobile robot based on edge-cloud collaborative computing. Int. J. Syst. Assur. Eng. Manag, **13**, 594–604 (2022). https://doi.org/10.1007/s13198-021-01545-6
2. Rajput, U., Kumari, M.: Mobile robot path planning with modified ant colony optimisation. Int. J. Bio-Inspired Comput. **9**(2), 106 (2017). https://doi.org/10.1504/IJBIC.2017.083133
3. Mamur, H., Üstüner, M.A., Bhuiyan, M.R.A.: Future perspective and current situation of maximum power point tracking methods in thermoelectric generators. Sustain. Energy Technol. Assess. **50**, 101824 (2022). https://doi.org/10.1016/j.seta.2021.101824
4. Bai, L., Du, C.: Design and simulation of a collision-free path planning algorithm for mobile robots based on improved ant colony optimization. Ingenierie des Systemes d'Information **24**(3), 331–336 (2019). https://doi.org/10.18280/isi.240313
5. Weiss, R.M.: GPU-accelerated ant colony optimization. In: GPU Computing Gems Emerald Edition, Elsevier, pp. 325–340 (2011). https://doi.org/10.1016/B978-0-12-384988-5.00022-X
6. Bi, Z., Ma, Y., Yang, X., Zhou, P.: Research on wireless robot path planning under edge computing considering multistep searching and inflection points. Trans. Emerg. Telecommun. Technol. **32**(6) (2021). https://doi.org/10.1002/ett.3825
7. Nayak, S., Patgiri, R., Waikhom, L., Ahmed, A.: A review on edge analytics: issues, challenges, opportunities, promises, future directions, and applications. Digit. Commun. Netw. **10**(3), 783–804 (2024). https://doi.org/10.1016/j.dcan.2022.10.016
8. Song, B., Tang, S., Li, Y.: A new path planning strategy integrating improved ACO and DWA algorithms for mobile robots in dynamic environments. Math. Biosci. Eng. **21**(2), 2189–2211 (2024). https://doi.org/10.3934/mbe.2024096

Application of System Dynamic Modelling in Mitigating Engineering Project Management Challenges

Daniel Harith Wong(✉), Christina Chin May May, and Novita Sakundarini

University of Nottingham Malaysia, 43500 Semenyih Selangor, Malaysia
wongdanielww@gmail.com

Abstract. This research investigated engineering PM challenges of large-scale oil and gas industry projects and applied System Dynamic Modelling (SDM) in mitigating these project challenges. In-depth interviews with experienced and qualified project managers and senior project engineers in oil and gas industry were conducted to gather information on PM challenges and mitigation plans. Based on the outcomes, the causes and effects of the PM challenges were identified, and mitigation plans were prepared as input to SDM to address the most common engineering challenges. Causal Loop Diagrams (CLDs) were prepared using Vensim to capture the mitigation plans for these most common engineering challenges, which include engineering project team selection and mobilization, and engineering scope changes. The results show that SDM can be a complementary PM tool to the existing traditional PM techniques and tools to mitigate the dynamic nature of PM challenges, especially for complex and large-scale projects in the industry. It captures a holistic view of PM challenges as well as the proactive mitigation plans by focusing on the interrelation and feedback processes within the project environment and boundary. However, SDM has yet to be adopted to mitigate PM challenges and its application is still relatively new for engineering PM, particularly in the oil and gas industry.

Keywords: Large Projects · Project Management · System Dynamic Modelling

1 Introduction

Project management (PM) is a complicated process, and even the most experienced project managers face numerous PM challenges over the course of the project to complete the project successfully. These include project constraints such as tight project schedules and budgets, demand for high quality from clients, and ensuring a motivated project team. PM of large-scale oil and gas projects is doubly complicated given the unique features of the industry, which is capital intensive, highly complex and requires state-of-art technology. Hence, despite the near-perfect project planning, availability of qualified project team members and PM tools, about 70% of projects still failed [1].

The traditional PM tools, such as the Work Breakdown Structure (WBS), Responsibility Matrix, and Critical Path Method (CPM) network, do not address PM challenges

Chin, C. M. M. et al. (Eds.): SSIEC 2024, PTT, pp. 57–64, 2025.
https://doi.org/10.1007/978-981-96-2806-3_10

exhaustively. Since the inter-relationship between the elements is complicated, a proper method/tool need to be used to understand better to completely circumvent the risks of the project [2, 3]. Here, SDM could be used to simulate PM challenges with the purpose of mitigating them. It is the objectives of this study to identify and investigate the challenges of PM in complex projects in the oil and gas industry, then utilize SDM to simulate PM challenges perspective, and address and circumvent the risks. Finally, to develop and document recommendations based on solutions derived from SDM applications to improve the management of complex projects.

2 Literature Review

2.1 PM and Challenges

PM is the series of activities to deliver a project to the client, meeting all the project objectives. The project manager must have adequate knowledge, skills, techniques, and tools, and able to lead a team to achieve the project goals within the constraints of scope, time, and budget [4]. To facilitate project planning and management, PM processes are generally broken down into project initiation, planning, execution, control and close-out [5]. Each phase will encapsulate many sub-processes for project managers to perform.

It is the project manager's responsibility to manage the project until its successful completion, but with many challenges. To successfully execute the project within budget and schedule while ensuring high quality, it is essential to conduct thorough planning and closely monitor tasks throughout all phases, thereby addressing potential challenges and risks. The causes of project failure vary from projects with companies spending vast amounts of money and time to address the PM challenges they encounter.

PM in the oil and gas industry is especially challenging as there are unique requirements for managing technology, science, and multi-discipline engineering aspects of projects. It is a complex, labor-intensive process with a long supply chain. In addition, clients' demand for high safety, integrity, and quality has added to the complexity of these projects. All these unique characteristics of the projects need a different technique in PM. Other key characteristic is the selection of project team individuals from different locations of the same company. In most international projects, these individuals vary from countries, educational backgrounds, cultures, employment and skills. With those differences, project manager must ensure they can work together to complete the project successfully within the budget and specific time [7]. Most of the large-scale oil and gas projects are either national or international projects. Hence, managing oil and gas projects successfully is essential for the national economic vitality of the country.

The top 5 most common engineering PM challenges identified are project team selection and mobilization [6, 8, 12], engineering interdisciplinary communication [6, 9, 15], engineering scope change [6, 7, 11], engineering quality management [6, 12, 16] and stakeholders management [6, 15].

2.2 System Dynamic Modelling (SDM)

SDM has been adopted as an effective analytical tool in certain situations and sector applications. This is because it captures, simulates, and analyses all the possible changes and feedback processes responsible for the subject's behavior and challenges. The process has five steps, (1) problem or challenge articulation, (2) formulation of dynamic hypothesis, (3) formulation of simulation model, (4) model testing and review and (5) policy and decision making. It is an iterative process where the developed model goes through constant modelling, testing, and refinement. Then, the modelling process is repeated until the simulated model's structure and behaviour reflect the dynamic reality of the problem or challenge [17]. SDM features a holistic simulation where the challenges are modelled and analysed as an entire system. Feedback loops are significant and vital in understanding the interrelationships between the fundamental variables [18, 19]. It also has a feedback analysis where the behaviour and structure of the interconnected elements are analysed for a solution [20].

3 Methodology

Candidates with more than 10 years of PM experience in engineering or Engineering Procurement and Construction (EPC) projects were sourced and selected for one-to-one interviews the questions cover the years of work experience of the interviewees, types of projects handled before, PM challenges encountered at each phase of the project and the types of PM tools adopted. In depth interviews, averaging about an hour, were conducted from Oct 2022 to Jan 2023. For SDM simulation to mitigate engineering PM challenges, data from literatures and interviews were analyzed. Simulated results of CLDs were prepared using Vensim, capturing and reflecting the mitigation plans for the challenges.

4 Results and Discussion

Interviewees agreed that there were numerous PM challenges despite comprehensive plannings made during the proposal and initial stage of the project. Numerous PM techniques and tools had been deployed, but many projects still could not meet clients' expectations. Per interview feedback, each company had their own respective PM tools, but the generic ones include:

- WBS to make the work more manageable and deliverable oriented for execution.
- Responsibility matrix to outline the roles of various stakeholders in completing project deliverables.
- Primavera with CPM to manage resources and schedule.
- Change request register to capture engineering scope changes.
- Standardized reports to communicate with stakeholders on project progress.

All the interviewees have yet to utilize SDM. However, they do not rule out the benefits of using SDM to mitigate PM challenges of large oil and gas projects. SDM simulation was performed for the top 5 common engineering PM challenges with C&Es, mitigation plans, and CLDs had been prepared. The CLDs capture the activity of the mitigation plans holistically to mitigate the challenges. The interviewees reviewed the results for accuracy and completeness and incorporated their comments. Two of the SDM results of mitigating engineering PM challenges are representatively appended herewith.

4.1 Engineering Project Team Selection and Mobilization Challenges

The success of an engineering project relies substantially on a qualified and experienced teamwork and effective communication skills. Selection of an engineering project team that meets the stringent requirements is crucial to prevent project failures. Mobilization of the team is in line with project planning and timing so the relevant engineering deliverables can be issued and carried out with precision timing. The C&Es and mitigation plans for team selection and mobilization challenges are in Table 1.

Table 1. C&E of engineering project team selection and mobilization

No.	Challenges	Causes & Effects	Mitigation Plans
1	Project team members not properly selected based on work requirements	Skills and experiences not fitting into the positions. Not able to perform the roles as needed.	Select the right candidates for the key project positions.
2	Unavailability of suitable skilled staff	Contractor does not have sufficient qualified staff. Not able to perform the work per project requirements.	Train staff and recruit skilled workers who fit into the job requirements.
3	Not able to form a complete project team	Sketchy project team members and not able to execute project properly.	To form a project team in accordance to project requirements.
4	Project team members are not mobilized timely	Too early mobilization will waste manhours. Too late mobilization will delay work progress.	Prepare manpower mobilization per project requirements.

Based on the C&Es and mitigation plans, the CLD is prepared and repeatedly simulated to capture all activities, as shown in Fig. 1. The main activities (red circle) and the respective sub-activities (blue arrow) are simulated to address project team selection and mobilisation challenges.

4.2 Engineering Scope Change Challenges

Over the course of engineering projects, client often initiates scope changes which can significantly affect the project cost and schedule. These changes are significant if it impacts the multi-discipline designs and procurement processes. Feedback from interviews revealed that some of these scope changes were not communicated effectively. Scope changes are made without clear client instructions and total input from all the relevant engineering disciplines. As a result, engineering scope change will impact the project cost, schedule, materials, pose additional risks, and jeopardize work quality if not properly managed.

For this research, the focus is on engineering scope change initiated by the client. C&Es and mitigation plans are tabulated in Table 2, and the activities to mitigate the challenges are captured in the CLD shown in Fig. 2. As captured in the CLD, which has been repeatedly simulated to capture all activities. The main activities (red circle) and the respective sub-activities (blue arrow) are conducted to mitigate the challenges.

Table 2. C&E of Engineering scope change challenges

No.	Engineering Scope Change Challenges	Causes & Effects	Mitigation Plans
1	Client can request for scope change at any time during project duration and at whatever situations	Project progress will be interrupted, progress delayed with budget impact etc.	Controlled situation and plan should be in place. Scope change procedures and execution plan to be set up and agreed by Client.
2	Scope change not properly reviewed by Contractor	Contractor misunderstands the scope and impacts not properly assessed. Details are missed out and project team has no full picture of the change.	Project manager calls for internal meeting with all relevant engineering disciplines to understand the scope and assess the impacts.
3	Scope change not properly prepared and not officially submitted by Contractor	Schedule and cost impacts are wrongly assessed and erroneously documented in the change order proposal.	Detailed and accurate change order proposal to be prepared for submission to Client for approval.
4	Scope change not approved by Client	Work proceeded but change proposal not approved by Client	Not to start change order work until Client has approved the change order proposal.
5	Scope change not properly executed	Change order work wrongly executed and need to be reworked.	Project team to be briefed clearly on the scope change.

4.3 SDM – A Practical PM Tool

The traditional PM tools such as Gantt chart, PERT diagram and CPM focus on the project work structure, scheduling and budgeting project activities, monitoring and controlling project performance. It assumes the perspective of "static and closed" of the project throughout the project duration. The reality is that the project is dynamic, changes are unavoidable, and they pose new challenges and risks to the project. Table 3 summarizes the key improvements of adopting SDM in mitigating the challenges.

Table 3. Comparison of Traditional and SDM Approaches

Traditional Approach		SDM Approach
WBS, Gantt Chart, PERT, CPM	Focus on monitoring & controlling based on initial project info.	Identify potential scope changes and challenges proactively.
Responsibility Matrix	Assign party to issue project deliverables and do not identify clearly the action party for challenges.	Assign action party to address challenges upfront.
Cost Estimation	Prepare project estimate but does not address the mitigation plan for overrun.	Address challenges upfront to avoid cost overrun.
Project Control: Varian analysis, earned value etc.	A reactive tool to detect project overruns and recommend corrective actions.	Provide holistic view to address challenges based on continuous feedbacks and proactive actions.

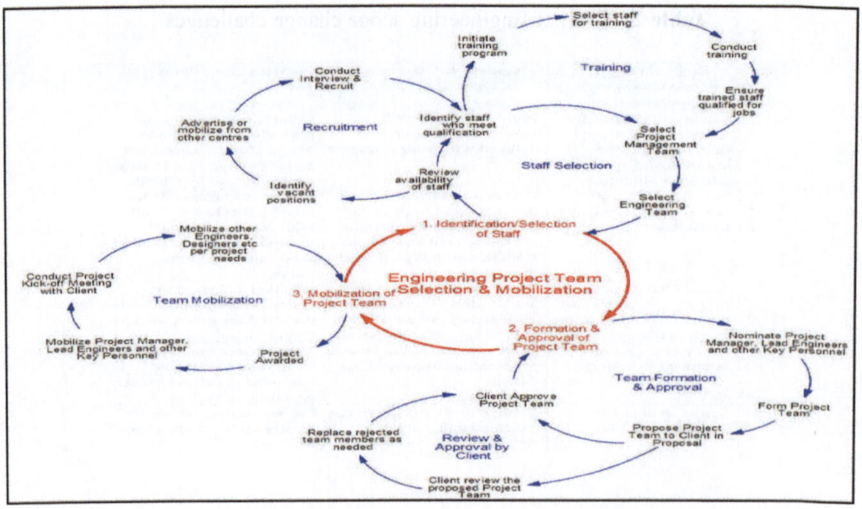

Fig. 1. CLD for Engineering project team election and mobilization

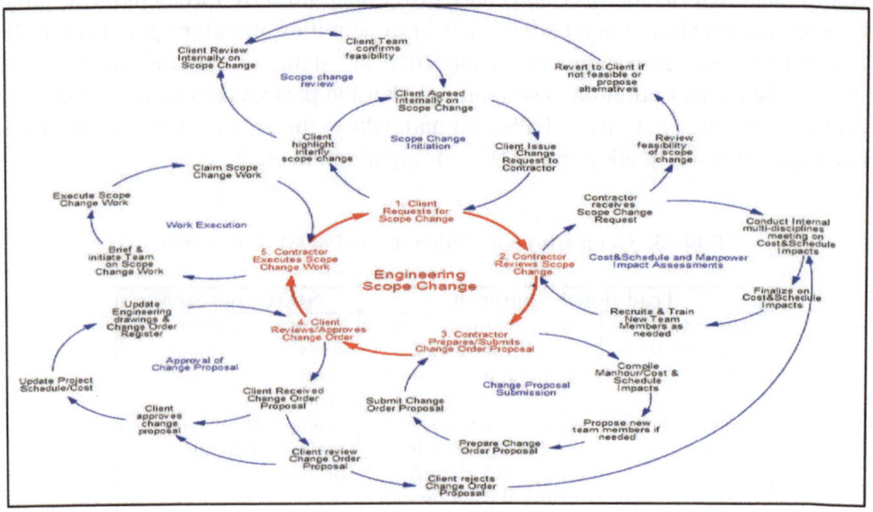

Fig. 2. CLD for engineering scope change

5 Conclusion

The study revealed that the traditional PM techniques and tools need to be improved to mitigate the dynamic nature of PM challenges, which resulted in many project failures. The challenges are more intriguing for complex and large-scale projects in the industry, and based on the feedback, SDM has yet to be adopted to mitigate PM challenges. SDM provides the complementary PM tool with a proactive engagement plan and focuses on the inter-relation and feedback processes within the project environment and boundary.

It emphasizes the inter-relationships of project elements that may causes of unexpected schedule delays and budget overrun. In this research, the results of SDM provided a holistic view of the complete project steps and activities to address and circumvent the challenges as well as the project risks and to develop mitigation plans based on solutions derived from SDM to improve the PM of complex projects. The application of SDM is still relatively new particularly in the oil and gas industry as they had yet been introduced to adopt SDM in mitigating PM challenges, reason viewed as taking additional time and effort to add SDM to the existing list of PM tools. However, there is vast potential in SDM and is a way forward for industry to try out in some of their projects to justify that SDM can be adopted as a useful tool in mitigating the challenges and improving PM overall. Additionally, SDM can be developed to facilitate the simulation process tailored for PM of large or complex projects. The long-term plan is to integrate SDM into the traditional PM tools in mitigating engineering PM challenges to ensure project success.

References

1. TeamStage. 31 Pivotal Project Management Statistics for 2021. TeamStage (2020). https://teamstage.io/project-management-statistics/
2. Rodrigues, A.G.: The application of system dynamics to project management an integrated methodology (n.d.). https://stax.strath.ac.uk/concern/theses/g158bh37n
3. Rodrigues, A.G., Williams, T.M.: System dynamics in Project Management: assessing the impacts of client behaviour on project performance. J. Oper. Res. Soc. **49**(1), 2–15 (1998). https://doi.org/10.1057/palgrave.jors.2600490
4. Heagney, J.: Fundamentals of Project Management. 6th edn. McGraw-Hill, New York (2022)
5. Project Management Institute: A guide to the Project Management Body of Knowledge: (PMBOK® guide). 6th edn. Project Management Institute (2017)
6. Borhanpour, M.: Main causes of Project failure. MBPMP (2015). https://mbpmp.wordpress.com/2015/01/05/5/
7. El-Reedy, M.A.: Project Management in the Oil and Gas industry. Wiley, Hoboken (2016)
8. Akhavan Tabassi, A., Bryde, D., Mustafa Kamal, E., Dowson, J.: Challenges for Project Management in the 21st Century (2019). https://doi.org/10.15405/epms.2019.12.63
9. Quinn, J.: What to do if your project runs over- budget, and how to prevent it. Project-Management.com (2012). https://project-management.com/what-to-do-if-your-project-runs-over-budget-and-how-to-prevent-it/
10. Kojic, M.: Project cost management: everything you need to know. Clockify Blog (2022). https://clockify.me/blog/business/project-cost-management/
11. Faust, B.: 9 Project Management Challenges and How to Overcome Them Effectively (2021). https://rindle.com/blog/10-project-management-challenges-and-how-to-overcome-them-effectively
12. Kendrick, T.: 101 Project Management Problems And How to Solve Them: Practical Advice for Handling Real-World Project Challenges. American Management Association (2011)
13. Grey, S.: Origins of Schedule Overrun. Paper presented at PMI® Global Congress 2010—Asia Pacific, Melbourne, Victoria, Australia. Project Management Institute (2010)
14. Dudley, G.: A practical approach to project management procedures. PM Netw. **15**(6), 51–53 (2001)
15. Wyrozębski, P., et al.: Project Management - Challenges and Research Results (2016). https://doi.org/10.6084/m9.figshare.4502792

16. Ray, S.: Project Quality Management - A Quick Guide. ProjectManager.com (2019). https://www.projectmanager.com/blog/project-
17. Sapiri, H., Zulkepli, J., Ahmad, N.: Introduction to System Dynamic Modelling and Vensim Software. Penerbit Universiti Utara Malaysia (2017)
18. Rodrigues, A., Bowers, J.: The role of system dynamics in project management. Int. J. Project Manag. **14**(4), 213–220 (1996). https://doi.org/10.1016/0263-7863(95)00075-5
19. Rodrigues, A., Bowers, J.: Syst. Dyn. Rev. **12**(2), 121–139 (1996). https://doi.org/10.1002/(SICI)10991727(199622)12:2%3C121::AID-SDR99%3E3.0.CO;2- X
20. Ford, D.N., Lyneis, J.M.: System dynamics applied to project management: a survey, assessment, and directions for future research. In: Dangerfield, B. (eds.) System Dynamics. Encyclopedia of Complexity and Systems Science Series. Springer, Cham (2020). https://doi.org/10.1007/978-1-4939-8790-0_658

Lithium Battery Recycling Process in Malaysia

Zhi Shan Wu(✉), Christina Chin May May, and Novita Sakundarini

University of Nottingham Malaysia, 43500 Semenyih Selangor, Malaysia
{efyzw3,Christina.Chin,Novita.Sakundarini}@nottingham.edu.my

Abstract. Electric vehicles are introduced to the Malaysian market due to the increase of energy demand in transportation sector. The batteries reaching end-of-life consists of harmful chemicals endangers wellbeing of society and ecosystem, moreover the materials contain in the battery are valuable. Hence, the idea of recycling used lithium-ion (Li-ion) battery has been introduced to conserve the environment and reusable substances in the used batteries. Firstly, it is important to carry out investigation on the composition, disassembly process and recycling methods of Li-ion batteries that is used in an electric vehicle. Next, identifying list of resources required in setting up a battery recycling factory helps to calculate and estimate the total cost and budgeting for the set-up of a battery recycling factory. The recycling factory was planned to operate with a 550 tonnes capacity annually. It was estimated that the recycling factory would initially generate a gross and net profit of RM 6,389,545.16 and RM 4,497,969.41 respectively. The recycling factory will require RM 4,745,284.67 of investment as a kick start of the business. By the end of this study, a budgeted profit and loss account, a 12-month cash flow forecast will be proposed according to the annual capacity of 550 tonnes.

Keywords: Lithium-ion Battery · Electric Vehicle · Recycling

1 Introduction

1.1 Background Problem

In Malaysia, the energy demand for transportation sector had risen, resulting being one of the largest contributors of carbon emission as their energy efficiency of road transportation was only 18.88% [1]. Hence, electric vehicles are introduced into market. As production rate of electric vehicle increases, production of lithium-ion (li-ion) battery is growing as well. With this trend, batteries reaching end-of-life subsequently brings environmental issue like toxic heavy metals and harmful chemicals that will endanger wellbeing of society and ecosystem [2].

This study aims to contribute for a healthier environment, it is significant to plan for the end-of-life of the li-ion batteries. When disposing the batteries, it brings negative impact to the environment if not handled in a cautious manner as it causes leaching of harmful components from the battery into the environment [2, 3]. Moreover, the raw materials and manufacturing cost needed to manufacture new li-ion batteries are expensive as resources are scarce, on top of that, extracting raw materials over the years

Chin, C. M. M. et al. (Eds.): SSIEC 2024, PTT, pp. 65–70, 2025.
https://doi.org/10.1007/978-981-96-2806-3_11

also brought harmful impact to the environment [3]. The disassembly process adopted in this study is chosen as it reduces waste and emission of harmful substances, it limits the usage of highly acidic chemicals, and all possible toxic breakdowns will be handled with caution, where the recycling process aims to identify reusable resources to reduce waste. A battery recycling plant layout will be designed according to the annual capacity of 550 tonnes, which is about 6180 units as each li-ion battery weights about 889 kg.

1.2 Aims and Objectives

This study aims to carry out a full investigation on electric vehicle battery recyclability in Malaysia, it needs to attain the following research objectives:

- Examining the composition of Li-ion battery.
- Investigating the disassembly process and recycling method of Li-ion battery.
- Estimating the cost and resources needed to set up a battery recycling plant in Malaysia.

2 Literature Review

2.1 Composition of Lithium-Ion Battery

Li-ion battery is a type of rechargeable battery uses the reversible reduction of Li-ions to store energy and chosen for operating an electric vehicle due to its high energy density, great lifespan, light weight, and high energy efficiency [4]. It contains 8 kg of Lithium, 35 kg of Nickel, 20 kg of Magnesium and 14kg of Cobalt. Within the containing cells, the Li-ions move from the negative electrode to the positive electrode passing through an electrolyte. During operation, Li-ion move from the negative electrode to the positive electrode during discharge and back when charging [5]. The Li-ion use the electrolyte as a transmission medium undergoing the process of diffusion through the separator which will be embed into the negative electrode to ensure the receiving of Li-ion. Whereas during discharge state, the process will be reversed, which the Li-ion will be released from the negative electrode and use the electrolyte as a transmission medium undergoing the process of diffusion through the separator and return to the positive electrode [6]. It consists of four main components namely, cathode, anode, electrolyte, and separator [7].

2.2 Disassembly Process of Lithium-Ion Battery

Disassembly process is an essential step for recovering and reusing the product. Complete disassembly is the common disassembly process by carrying out total dismantling of the battery, all the components will need to be separated [8]. Although this could be an effective and useful method, but the cost and resources incurred is high. However, most of the research were carried out by examining the complete disassembly process [8, 9]. To introduce a safe and cost-effective disassembly process, firstly, it is important to discharge the battery till 0% of the state of charge, this process can be carried out effectively using a resistor or short circuit [10]. Next, it is important to check the voltage with a multimeter before cutting the battery pouch, this process should be carried out in a fume hood and preventing the use of metallic tools throughout as there might be risk of electric discharge [11]. Next, pouch cells will be undergoing separation process, it is optimum to separate the cell components into anode, cathode, and other materials as it prevents the risk of cross contamination during the process [4, 8]. Next, the electrode, also known as 'black mass' is now able to be separated from the copper and aluminium foil sheets. For the discharging of the materials from the current collector, using heat to break down the binder, solvent delamination or using ultrasonic will be an effective method [8, 9]. Finally, the delamination process of the anode and cathode will be separated from the copper and aluminium sheets in the cells.

2.3 Recycling Process of Lithium-Ion Battery

The recycling methods identified includes pyrometallurgical recovery, physical materials separation, hydrometallurgical metals reclamation and direct recycling [9]. However, this study focuses on pyrometallurgical recovery method as it provides a high reaction rate by using a high temperature furnace to break down the component metal oxides to an alloy of cobalt, copper, iron and nickel. The use of high temperature is needed as the batteries need to be heated and melted [12]. With the use of metal current collectors throughout the heating and melting process, this method has an advantage which can be used with whole cells or modules, without requiring coating process. At higher temperatures the polymers decompose and burn off. The metal alloy can be separated through hydrometallurgical processes into the component metals and the waste typically contains the metals aluminum, manganese, and lithium, which can be reclaimed by further hydrometallurgical processing. In the pyrometallurgical process, there isn't consideration given for reclamation of electrolytes and plastics or other components such as lithium salts [1].

3 Methodology

Milestone 1: This study was intended to carry out field study in the form of an interview or factory visit. However, the field study was unable to carry out as there were either no respond from stakeholders or the respective factories responded that they are not dealing with disassembly of batteries. Hence, the study will be continued on literature review based.

Milestone 2: The machineries and equipment used are determined according to the disassembly and recycling process reviewed. In the market, manufacturers are offering the Li-ion battery recycling machine. However, for this research it is suitable to select individual equipment according to the recycling process. The machine's power consumption was listed for estimating cost on electricity and utilities. The location of the factory has been selected by considering many factors such as the ability to get manpower, the rental and the area of the factory. The size of the factory was estimated according to the size of the machines and the factory layout plan, these was determined by the annual capacity of the factory. The estimated manpower required, and their wages were calculated according to the average worker's salary in Malaysia.

Milestone 3: In the results and discussion section, a budgeted profit and loss account for 2 years and cash flow forecast for the first year has been presented, Microsoft Excel was the tool used to create the projection. It estimates all form of expenses and assets the factory would incur, where the pricing and rates were estimated based on the current selling or market price.

4 Results and Discussion

In the first year, it is projected that the factory will be dealing with 3,090 units of used batteries and subsequently increasing till 6,180 units in the following year. Each Li-ion battery weighs about 889 kg, hence, it is estimated that the factory can deal with about 6180 units of used batteries [13]. Each refurbished batteries have a market price of RM13,716.68; it is estimated that the factory could generate sales revenue of RM42,384,541.20 and RM84,769,082.40 respectively. It is calculated that each new and used Li-ion battery worth about RM49,335 and RM8,706.18 respectively [14]. The total cost of sales for year 1 and year 2 shall be RM35,994,996.04 and RM71,240,092.24 respectively. It generates a gross profit of RM6,389,545.16 and RM13,528,990.16 respectively; gross profit margin of 15% and 16% respectively. The total overhead cost of both year 1 and year 2 were estimated to be the same amounting to RM1,580,375 as the overhead cost every year will not be affected by production volume. The net profit of this recycling factory in year 1 and year 2 were estimated to be RM4,497,969.41 and RM11,637,414.41 respectively.

The 12-month cash flow forecast for year 1 is projected according to the monthly production or sales quantity which is 257 units along with a selling price of RM13,716.68 is presented. The initial cost of sales presented for each particular will be the amount estimated at the start of the business and amount incurred per month over the 12-month period where both total amounting to RM2,999,583. The total overhead cost incurred at the start of the business and the total overhead cost incurred per month amounts to RM189,697.92 and RM131,697.92 respectively. The total cost of all machinery and equipment amounted to RM1,556,003.75. This 12-month cash flow forecast concludes that the battery recycling business will require RM4,745,284.67 of investment for a kick start.

5 Conclusion

This study suggested that:

- To operate the recycling factory, it is required to have a Li-ion battery charger and discharger machine for discharging the used batteries, a Li-ion battery recycling shredder to tear large used batteries, a filter press for separation of black mass, a battery separator machine which separates various materials, an evaporation and heated tank system to evaporate excess water and concentrating the brine, and a tail gas treatment system for gas treatment processes.
- It is required to have a 15,000 sq.ft. land area factory to recycle capacity of 550 tonnes annually.

- By generating the budgeted profit and loss, the budgeted selling price of refurbished Li-ion battery was decided to be at RM13,716.68 and the projected sales quantity for year 1 and year 2 shall be 3,090 and 6,180 respectively. Therefore, this results in the sales revenue for each year to be RM42,384,541.20 and RM 84,769,082.49 respectively.
- The budgeted P&L suggested that the recycling factory for each year would generate a gross profit and gross profit margin of RM6,389,545.16 and 15%; RM13,528,990.16 and 16% respectively. Whereas for net profit, it generates an amount of RM4,497,969.41 and RM11,637,414.41 respectively.
- The cash flow forecast suggested that RM4,745,284.67 of investment is required to set up the recycling factory and would have a return on investment of 0.95 and 2.45 for year 1 and year 2 respectively.

References

1. Costa, C.M., Barbosa, J.C., Goncalves, R., Castro, H., Campo, F.J.D., Lanceros-Mendez, S.: Recycling and environmental issues of lithium-ion batteries: advances, challenges and opportunities. Energy Storage Mater. **37**, 433–465 (2021)
2. Park, J.S., Seo, S., Han, K., Lee, S., Kim, M.J.: A process using a thermal reduction for producing the battery grade lithium hydroxide from wasted black powder generated by cathode active materials manufacturing. J. Hazardous Mater. (2023)
3. Harper, G.D.J., Kendrick, E., Anderson, P.A., Mahanty, S., Boons, F.: Roadmap for a sustainable circular economy in lithium-ion and future battery technologies. J. Phys. Energy (2023)
4. Hendricks, C.E., Mansour, A.N., Fuentevilla, D.A., Waller, G.H., Ko, J.K., Pecht, M.G.: Copper dissolution in overdischarged lithium-ion cells: x-ray photoelectron spectroscopy and x-ray absorption fine structure analysis. J. Electrochem. Soc. **167**, 090501 (2020)
5. Kalpana, R., Nagde, S.J.D.: Chapter 12 - Li-S ion batteries: a substitute for Li-ion storage batteries. In: Dhoble, S.J., Thejo Kalyani, N., Vengadaesvaran, B., Arof, A.K. (eds.) Energy Materials, pp. 335–371. Elsevier, Amsterdam (2021). ISBN 9780128237106, https://doi.org/10.1016/B978-0-12-823710-6.00008-X. https://www.sciencedirect.com/science/article/pii/B978012823710600008X
6. Shaw-Stewart, J., et al.: Aqueous solution discharge of cylindrical lithium-ion cells. Sustain. Mater. Technol. **22**, e00110 (2019)
7. The Four Components of a Li-ion Battery. Samsung SDI (2016). https://www.samsungsdi.com/column/technology/detail/55272.html#:~:text=Li%2Dion%20batteries%20consist%20of,of%20the%20components%20is%20missing. Accessed 7 Oct 2022
8. Williard, N., Sood, B., Osterman, M., Pecht, M.: Disassembly methodology for conducting failure analysis on lithium-ion batteries. J. Mater. Sci. Mater. Electron. Electrochim. Acta **22**, 1616–1630 (2011)
9. Diekmann, J.: Ecological recycling of lithium-ion batteries from electric vehicles with focus on mechanical processes. J. Electrochem. Soc. **164**, A6184–A6191 (2017)
10. Marshall, J., Gastol, D., Sommerville, R., Middleton, B., Goodship, V., Kendrick, E.: Disassembly of Li Ion Cells—Characterization and Safety Considerations of a Recycling Scheme, vol. 1, p. 5 (2020)
11. Meshram, P., Pandey, B.D., Mankhand, T.R.: Extraction of lithium from primary and secondary sources by pre-treatment, leaching and separation: a comprehensive review. Hydrometallurgy **150**, 192–208 (2014)

12. Wegener, K., Chen, W.H., Dietrich, F., Dröder, K. Kara, S.: Robot assisted disassembly for the recycling of electric vehicle batteries. Proc. CIRP **29**, 716–721 (2015)
13. The Weight Of A Battery Can Vary: How Much Does An EV Battery Weigh. AXLEWISE (2023). https://axlewise.com/electric-car-battery-weight/. Accessed 18 Mar 2023
14. Used Lithium battery. Jiangsu GSO New Energy Technology Co., Ltd. Custom manufacturer (2022). https://www.alibaba.com/product-detail/OEM-Widely-Used-48V-Lithium-Ion_1600197636858.html?spm=a2700.pc_countrysearch.main07.162.6c2a6570lvBblH. Accessed 18 Mar 2023

“Doing Well by Doing Good”: How Does Socially Responsible Investing Influence Innovation?

Yi Ke and Shu-Hsiang Chen(✉)

Entrepreneurship and Enterprise Hub, Entrepreneur College (Taicang), Xi’an Jiaotong-Liverpool University (XJTLU), Suzhou, China
ShuHsiang.Chen@xjtlu.edu.cn

Abstract. This research investigates the relationship between Chinese energy firms’ socially responsible investing (SRI) and their innovation. Based on the stakeholder theory and recent research on firms’ innovation and sustainability, we conceptualize that SRI uncertainly influences a firm’s innovation, and this mechanism depends on the time effects derived from the firm’s strategic corporate social responsibility (SCSR) theory. The research plan to leverage the data from Chinese listed energy firms spanning ten years and regression year by year. Additionally, the research set international leading firms as benchmarking firms in this research. These benchmarking exercises are envisaged to yield instructive insights, thereby furnishing more efficacious strategies derived from firms that epitomize best practices. In a short-time period, SRI decreases a firm’s innovation; however, in a long-term period, SRI could boost innovation. This research aims to bridge the gap between stakeholder theory and SCSR theory in innovation and sustainability research and provide applicable suggestions for Chinese energy firms.

Keywords: Socially Responsible Investing · Innovation · Energy Firms

1 Introduction

According to the National Bureau of Statistics of China, total energy consumption in 2023 was 5.41 billion tons of standard coal, an increase of 2.9% over the previous year. It is alarming that, without accelerating change in clean energy, China will lock itself into a carbon-intensive energy system. Innovation in clean energy technology is a critical backbone of the energy transition (Lin and Zhu, 2019). To change the current situation, the Chinese government has adopted some clean energy innovation strategies and measures (Liu and Liang, 2013). At the end of 2023, the cumulative power generation of nuclear power units accounted for 4.86% of China’s cumulative power generation, an increase of 3.98% over the same period in 2022. Hence, it is essential to focus on the factors that may boost innovation related to clean energy.

According to the World Bank Group, firms located in more than 200 countries have started to take social responsibility into account and are bearing socially responsible

Chin, C. M. M. et al. (Eds.): SSIEC 2024, PTT, pp. 71–75, 2025.
https://doi.org/10.1007/978-981-96-2806-3_12

investing (SRI). SRI can be defined as an investment process that considers both positive and negative social and environmental consequences within the context of rigorous financial analysis (Schwartz and Carroll, 2003; Social et al., 2006). Increasingly, organizations are concerned about how their actions affect the environment and social welfare (Sprinkle and Maines, 2010). Employees, consumers, investors, lenders, governmental agencies, and other stakeholder groups demand that firms operate socially responsibly (Jamali and Mirsha, 2007). As for the Chinese energy industry, firms with socially responsible investing are expected to operate in a 'smart' way (Lin and Zhu, 2019) and can achieve this by increasing their green innovation; however, the mechanism of how the Chinese energy firms' SRI influence their innovation has not yet been empirically investigated.

Studies have debated whether a firm engaged in SRI can lead to an increase in innovation. Innovation output usually refers to the creation of new products and new processes (Kleinknecht, 1999). Traditional economic theory (Friedman, 1970; Jensen and Meckling, 2019) highlights the importance of maximizing shareholder wealth. The increase in agency costs from SRI can place firms at an economic disadvantage and benefit other stakeholders at the expense of shareholders (Vance, 1975; Aupperle et al., 1985; Pagano and Volpin, 2005; Surroca and Tribó, 2008; Cronqvist et al., 2009; Kruger, 2015). A firm with an SRI might have less capital invested in innovation, and according to traditional economic theory, SRI is viewed as an expense. Thus, SRI can have a negative effect on a firm's innovation. In contrast, the enlightened stakeholder theory, as described by Jensen (2001) and Porter (2011), suggests that various stakeholders (e.g., shareholders, employees, customers, suppliers, the environment, and the community) play critical roles in supporting a firm and its operations through corporate social responsibility (CSR) activities, which ultimately contributes to shareholder value (Edmans, 2011; Deng et al., 2013; Servaes and Tamayo, 2013; Guiso et al., 2015). When the benefit is developed from a properly implemented strategic CSR program, it is recognized as an investment for the company, not an expense (Asongu, 2007). Recent studies show that firms with higher SRI generate more patents and patent citations, making them more likely to commit to a long-term focus on innovation (Cook et al., 2015). Therefore, according to the enlightened stakeholder theory, SRI is viewed as an investment contributing to a firm's innovation.

From the perspective of competitive strategy, Porter (2008) points out that it will be a win-win strategy for firms to seek their interest and consider society's interests. With the development of the "strategic corporate social responsibility" (SCSR) theory, corporate social responsibility has been recognized at the strategic level of a firm's competitiveness. Researchers believed that SCSR creates new business opportunities (Prahalad and Ramaswamy, 2004; Shao and Meng, 2011) and enables firms to establish an effective communication mechanism with stakeholders (Donaldson and Preston, 1995). Hence, according to Porter's SCSR theory, CSR is a fundamental strategic competitive tool. However, this theory did not take time related effects into consideration.

Time-dependent effects also influence the impact of SRI on a firm's innovation. In the short term, SRI might be viewed as an expense, leading to decreased innovation output because firms with SRI have less capital invested in innovation (Lougee and Wallace, 2008; Sarkar and Searcy, 2016). In the long term, SRI might be recognized as an investment which positively impacts a firm's innovation because SRI contributes to shareholder value; thus, firms have more capital for innovation investment (Deng et al., 2013; Carroll, 2021). Nevertheless, previous research should have accounted for the time effect in their frameworks (Cronqvist et al., 2009; Kruger, 2015; Servaes and Tamayo, 2013; Guiso et al., 2015; Ullah & Sun, 2021). Uncertainties persist regarding the temporal dynamics governing the engagement in social activities and the realization of tangible returns stemming from SRI initiatives. The proposed research endeavours to illuminate this enigma by scrutinizing the presence of a temporal inflexion point.

As SRI may negatively impact innovation in the short term, firms must deliver a long-term strategy and avoid short-sighted behaviour. Energy firms, in particular need to adjust their innovation policy from a long-term perspective to gain sustainable competitive advantages. Therefore, another objective of this research is to fill this gap by considering the time dimension, namely, short-term and long-term effects, and test how SRI drives a firm's innovation, which is expected to advance the theoretical development of SCSR theory.

2 Method

The dependent variable is SRI, which can be directly found in Chinese-listed energy firms' Corporate Social Responsibility (CSR) reports. This research adopts the natural logarithm of SRI numeric value plus one in the modelling to avoid some firms whose SRI equals zero. In this study, innovation is the independent variable; it simplifies the energy firm's innovation as the number of its added patents in that year. The data can be obtained from the SIPO (the State Intellectual Property Office) website.

To test the effects related to the time period, the research conducted a year-by-year regression, setting sample firms' age and total assets as control variables, then comparing the leading coefficient in each year, further finding whether the temporal turning point exists. Data cleaning and hypotheses testing will be conducted using the statistical software Stata.

3 Results and Discussion

The research predict that the regression coefficients in the past five years are negative, which explains that SRI decreases innovation as it may not generate an immediate return. However, in the last five years, the regression coefficients between SRI and innovation are positive, which means a temporal turning point exists. Moreover, the later years' coefficient has been more significant in the last five years. Thus, it further concludes that the longer the firms bear SRI, the more positive the impact on firms' innovation is.

When comparing the results with these benchmarking energy firms, the research found that the coefficient between SRI and innovation is more significant than that in Chinese energy firms. This phenomenon is because benchmarking firms undertake SRI for a longer time than Chinese energy firms, especially those private-fund energy firms. As SRI will not bring tangible benefits quickly, these benchmarking firms persist until SRI generates visible benefits. This practice requires the firm's solid financial backgrounds.

4 Conclusion

An important finding is the identification of a temporal inflexion point. In the short term, SRI may have a negative effect on firm innovation and, consequently, decrease clean energy share. However, in the long term, SRI is expected to impact firm innovation positively, leading to an increase in the share of clean energy.

This study also seeks to extend the understanding of firms' innovation strategies within the framework of energy sustainability, with a specific focus on the temporal dimension. While Strategic Corporate Social Responsibility (SCSR) has gained prominence, primarily due to Porter's seminal work in 2008, the temporal aspects associated with CSR and SCSR have received relatively limited attention within the existing literature. This research project boldly ventures into this uncharted territory by investigating the evolving impact of SRI on firms' innovation over time. This research promises to be a trailblazer in augmenting the SCSR theory by examining whether an inflexion point exists. By delving into the time-dependent aspects of CSR and SCSR, this study will pave the way for a more nuanced and comprehensive understanding of how firms strategically leverage CSR to foster innovation in the context of energy sustainability.

Regarding the managerial implication, this research underscores the importance of firms adopting a long-term perspective in their investment and innovation strategies, discouraging short-sighted behaviour and advocating for strategic planning over extended periods.

As for the future research directions, the envisaged applicability of the proposed research framework transcends the energy sector and extends its reach across a multiplicity of industries, prominently including the high-technology domain. This application seeks to contribute to the body of knowledge in clean energy adoption and sustainability while underscoring its potential applicability in various industrial contexts.

References

Lin, B., Zhu, J.: Determinants of renewable energy technological innovation in China under CO2 emissions constraint. J. Environ. Manag. **247**, 662–671 (2019)

Lin, B., Zhu, J.: Energy and carbon intensity in China during the urbanization and industrialization process: a panel VAR approach. J. Clean. Prod. **168**, 780–790 (2017)

Liu, H., Liang, D.: A review of clean energy innovation and technology transfer in China. Renew. Sustain. Energy Rev. **18**, 486–498 (2013)

Schwartz, M.S., Carroll, A.B.: Corporate social responsibility: a three-domain approach. Bus. Ethics Q. **13**(4), 503–530 (2003)

Social Investment Forum. Report on socially responsible investing trends in the United States: 10-year review (2006)

Sprinkle, G.B., Maines, L.A.: The benefits and costs of corporate social responsibility. Bus. Horiz. **53**(5), 445–453 (2010)

Jamali, D., Mirsha, R.: Corporate social responsibility (CSR): theory and practice in a developing country context. J. Bus. Ethics **72**(3), 243–262 (2007)

Brouwer, E., Kleinknecht, A.: Innovative output, and a firm's propensity to patent: an exploration of CIS micro data. Res. Policy **28**(6), 615–624 (1999)

Friedman, M.: The social responsibility of business is to increase its profits. Corporate ethics and corporate governance, pp. 173–178 (1970)

Jensen, M.C., Meckling, W.H.: Theory of the firm: managerial behavior, agency costs and ownership structure. Corporate Governance, pp. 77–132 (2019)

Vance, S.C.: Are socially responsible corporations' good investment risks? Manag. Rev. **64**(8), 19–24 (1975)

Aupperle, K.E., Carroll, A.B., Hatfield, J.D.: An empirical examination of the relationship between corporate social responsibility and profitability. Acad. Manag. J. **28**(2), 446–463 (1985)

Pagano, M., Volpin, P.F.: The political economy of corporate governance. Am. Econ. Rev. **95**, 1005–1030 (2005)

Surroca, J., Tribó., J.A.: Managerial entrenchment and corporate social performance. J. Bus. Finance Account. **35** 5–6), 748–789 (2008)

Cronqvist, H., Heyman, F., Nilsson, M., Svaleryd, H., Vlachos, J.: Do entrenched managers pay their workers more? J. Finance **64**(1), 309–339 (2009)

Kruger, P.: Corporate goodness and shareholder wealth. J. Financ. Econ. **115**(2), 304–329 (2015)

Edmans, A.: Does the stock market fully value intangibles? Employee satisfaction and equity prices. J. Financ. Econ. **101**(3), 621–640 (2011)

Servaes, H., Tamayo, A.: The impact of corporate social responsibility on firm value: the role of customer awareness. Manag. Sci. **59**(5), 1045–1061 (2013)

Guiso, L., Sapienza, P., Zingales, L.: The value of corporate culture. J. Financ. Econ. **117**, 60–76 (2015)

Asongu, J.J.: Innovation as an argument for corporate social responsibility. J. Bus. Publ. Policy **1**(3), 1–21 (2007)

Cook, K.A., Romi, A.M., Sánchez, D., Sanchez, J.M.: The influence of corporate social responsibility on investment efficiency and innovation. J. Bus. Financ. Acc. **46**(3–4), 494–537 (2015)

Porter, M.E.: On Competition. Harvard Business Press (2008)

Prahalad, C.K., Ramaswamy, V.: The future of competition: co-creating unique value with customers. Harvard Business Press (2004)

Donaldson, T., Preston, L.E.: The stakeholder theory of the corporation: concepts, evidence, and implications. Acad. Manag. Rev. **20**(1), 65–91 (1995)

Lougee, B., Wallace, J.: The corporate social responsibility (CSR) trend. J. Appl. Corp. Financ. **20**(1), 96–108 (2008)

Sarkar, S., Searcy, C.: Zeitgeist or chameleon? A quantitative analysis of CSR definitions. J. Clean. Prod. (135), 1423–1435 (2016)

Carroll, A.B.: Corporate social responsibility: perspectives on the CSR construct's development and future. Bus. Soc. **60**(6), 1258–1278 (2021)

Ullah, S., Sun, D.: Corporate social responsibility corporate innovation: a cross-country study of developing countries. Corp. Soc. Responsib. Environ. Manag. **28**(3), 1066–1077 (2021)

Author Index

Chin, C. M. M. et al. (Eds.): SSIEC 2024, PTT, p. 77, 2025.
https://doi.org/10.1007/978-981-96-2806-3

MIX
Papier aus verantwortungsvollen Quellen
Paper from responsible sources
FSC® C105338

If you have any concerns about our products,
you can contact us on
ProductSafety@springernature.com

In case Publisher is established outside the EU,
the EU authorized representative is:
Springer Nature Customer Service Center GmbH
Europaplatz 3, 69115 Heidelberg, Germany

Printed by Libri Plureos GmbH
in Hamburg, Germany